Synthesis Lectures on Solid State Materials and Devices

This series publishes short books with two basic goals: 1) develop the basic physical concepts to understand current and future electronic materials and devices; and 2) provide a sound understanding of current solid state devices and technology so that their applications to electronic circuits and systems can be appreciated. The series provides an overview of the most important types of solid state devices. Lectures will describe the operation and design issues of the devices, point out how materials properties affect device operation, and discuss the overall architecture issues for systems.

Rashad Gabil Oglu Abaszade · A. K. Haghi

Carbon Nanotube Devices for Nanoelectronics

Springer

Rashad Gabil Oglu Abaszade
Azerbaijan State Oil and Industry University
Baku, Azerbaijan

A. K. Haghi
Edinburgh, UK

ISSN 1932-1260 ISSN 1932-1724 (electronic)
Synthesis Lectures on Solid State Materials and Devices
ISBN 978-3-032-02113-7 ISBN 978-3-032-02114-4 (eBook)
https://doi.org/10.1007/978-3-032-02114-4

This Springer imprint is published by the registered company Springer Nature Switzerland AG
The registered company address is: Gewerbestrasse 11, 6330 Cham, Switzerland

Preface I

The twenty-first century has been characterized by a sharp increase in worldwide demands of energy simply because the world is experiencing an alarming rate of industrialization, population increase, and advancement in technology. The twenty-first century has been characterized by a sharp increase in worldwide demands of energy simply because the world is experiencing an alarming rate of industrialization, population increase, and advancement in technology. Combined with this has been the rising global environmental issues, namely; climate change, loss of resources, and pollution that have made the development of sustainable, clean and robust energy systems all the more pressing. In that regard, nanotechnology has not only become a complementary method, but a revolutionary technology that can redraw the basis of modern energy infrastructure.

The book *Carbon Nanotube Devices of Nanoelectronics* embarks on a bold and opportune vision into the world of nanoengineered materials, with a special consideration of carbon nanotube and its utilization in the next generation energy systems. It not only proposes an overview which is mainly technical, but a much broader and comprehensive view of the world including methods of synthesis, fabrics technique, an environmental standpoint, and the direction of energy related nanotechnology. Due to their out-of-this-world electrical conductivity, strength, thermal stability, and nanoscopic geometries, carbon nanotubes (CNTs) have the potential to overcome most of the drawbacks that existing energy systems have. As energy transition, this can either be the improvement of the capacity and charging rate of the lithium-ion battery, conversion efficiency in photovoltaic systems, or making it possible to store, and efficiently utilize hydrogen, CNTs are actively explored as the facilitators of the transition.

The book is designed so that it would take the reader through multiple thematic areas:

Chapter 1 has presented an introductory discussion on the synthesis processes of carbon nanomaterials, how they are integrated in electronic systems and how they are utilized in design of energy storage and conversion devices of a sustainable nature;

Chapter 2 redirects the interest to the ecological aspect where the environmental implication and life-cycle trace of the carbon nanostructures, especially those in relation to fuel cell technology have been assessed;

Chapter 3 includes a painstaking study of the inclusion of advanced nanomaterials in lithium-ion batteries, showing how the nanoscale innovations are playing a role in the higher energy density, stability and charge/discharge dynamics;

Chapter 4 addresses the topic of nanoengineered solutions in photovoltaic systems, in the case of solar cells, the architecture of the system, efficiency and reliability of CNT-based solar devices.

All through the book, technical performance is not the only factor which is put in individual light, but also the larger issues of commercial scalability, materials durability, recyclability and environmental compatibility are addressed. By so doing, it becomes a piece of work that has connected the worlds of basic nano science and applied energy engineering that are at times very separate. The dream at the root of this book is anchored in the ideology that, to achieve sustainable growth, we must be prepared to combine cross-discipline knowledge (materials chemistry to energy policy) and to be creative in how we think of the materials that will comprise our energy future. Carbon Nanotube Devices for Nanoelectronics is aimed at both researchers, graduate students, the working engineer and anyone interested in this synergy of nanotechnology and clean energy. It is an invitation and an appeal to the budding horizon of nanoenabled energy solutions, to critically involve in the move towards an energy future, not only efficient, high performance, but more importantly responsible and environmentally friendly.

We hope that this book will become a source of inspiration, of effective international partnerships, and of workable innovations in energy technologies that will allow the achievement of a sustainable and inclusive world.

Baku, Azerbaijan — Assoc. Prof. Dr. Rashad Gabil Oglu Abaszade
Edinburgh, UK — Assoc. Prof. Dr. A. K. Haghi

Preface II

Nanoelectronics leads the field of contemporary technological advances that allow an unrivaled advance in computation, communication, sensing and energy systems. Carbon nanomaterials are of high significance to the miniaturization of electronic materials and enhancing their effectiveness. This kind of paradigm shift has triggered the necessity to find out new materials capable of meeting the challenging requirement of the next generation electronic system.

One of the most promising material in this endeavor is CNTs classified as one of the one-dimensional nanostructures with mind-blowing electrical, mechanical, and thermal accomplishments. Carbon nanotubes represent a unique intercalation of properties since they appeared on the radar as a building block of nanotechnology seeming to tickle the fancy of both scientists and engineers alike. They also have big carrier mobility, bandgap tunable characteristics and band strength to weight ratio coupled with record conductive thermal properties and this offers numerous possibilities in new technology generation.

The current book entitled *Carbon Nanotube Devices for Nanoelectronics* has been conceived with the consideration of having a single comprehensive and yet easy to understand reference book that can be used by researchers, graduate students and also professionals operating in the field of nanotechnology and electronic materials. It brings together the foundation, the engineering properties and the emerging uses of CNTs in nanoelectronics.

The motivation behind putting together such a volume has been the rapid pace of research and consequential acceleration of volume of knowledge accreting around CNT-based devices. Whilst numerous investigations have been conducted into the individual areas of CNT synthesis, characterization and application, there is no yet single resource which ties the science behind these materials with the experimental engineering of devices. The book tries to fill that gap by giving a systematic account on CNT classification, properties, applications and the future trend.

In this study, special attention is devoted to how CNTs are applied in real-life electronic technologies, such as emission displays, digital storage, nanoscale electromechanical components, supercapacitors, and innovative energy retention systems. The trends of future research that were found in the book on CNTs include CNT-based sensors, green synthesis and environmental applications.

We hope that this work would be useful and inspirational to the scientific community. Whether you are a confirmed researcher attempting to expand your knowledge in the field of CNT-based electronics, a graduate student entering the latest frontiers of nanotechnology, or an engineer, the book is likely to provide you with a few ideas that will both inform and empower your activities. Without their help this book would have never seen the light of day. We would also like to thank our respective families and institutions that have been with us throughout in the compilation of this volume.

With our book Carbon Nanotube Devices for Nanoelectronics we wish to contribute perhaps a small part in the continuing evolution of nanoelectronics and to spur continued discovery in this most interesting and rapidly evolving field.

Baku, Azerbaijan — Assoc. Prof. Dr. Rashad Gabil Oglu Abaszade
Edinburgh, UK — Assoc. Prof. Dr. A. K. Haghi

Acknowledgments The journey towards preparing Carbon Nanotube Devices towards Nanoelectronics has been a rewarding as well as a collective experience and it could not have materialized without the untold aid and assistance of a multitude of individuals and institutions.

Ingredients to the coming of this book owe a lot to the pioneering works of scholars in the world who have yet to impact the future of nanotechnology through their research on carbon nanotube and nanoelectronics. Their experiences, findings and devotions are also resourceful and contribute to more innovations in this beautiful sphere.

We felicitate in particular our colleagues and co-workers, whose conversations, remarks and communicated learning have to a large measure filled these pages. Their frankness to exchange ideas and evolve an improved understanding of the subject has been instrumental in evolvement of this work.

We also wish to recognize the role played by our academic institutions and research centers whereby it has made us have the resources and the drive to work in an environment that is conducive to scientific research and academic writing. A well-established transportation network, along with advancements in smart societal systems, has had a profound impact on the effective implementation of this initiative.

Those editorial and publishing teams who have contributed towards the bringing of this book into reality are specially acknowledged. Their work ethics, diligence and issues of quality have ensured that the final product has been of the highest of academic standards.

Firstly, on individual grounds, we would wish to give our deepest thanks to our families who stood by us without any perturbation, were patient with us, and motivated us in the process of writing and compiling this book. Through the numerous hours of research and writing, they have served as an eye opener and a motivation to me.

And the last but not the least we would like to thank all those who read this book. Our forecast is that Nanoelectronics Carbon Nanotube Devices will be handy and will drive advances even in the future years. The further progress of nanoelectronics and the contribution of the carbon nanotube devices into the future technologies will be awaited by us greatly.

Assoc. Prof. Dr. Rashad Gabil Oglu Abaszade
Assoc. Prof. Dr. A. K. Haghi

Competing Interests The authors have no competing interests to declare that are relevant to the content of this manuscript.

About This Book

The book *Carbon Nanotube Devices for Nanoelectronics.*

The main idea of writing this book is to conduct a study and delve into the development and usage of nanoengineered materials in the growth of green energy frameworks and technologies. The need to utilize sustainable and renewable energy rises since the demand of energy in the world is increasing. With nanotechnology, it is now possible to address the problem of traditional energy system through the superior electrical, thermal, and mechanical materials. Any person who is fascinated by the field of materials science, nanotechnology, chemistry, energy engineering, as well as environmental science can achieve extensive knowledge concerning the field in this book. The book attempts to find out how nano-designed material utilization could improve performance and sustainability of the current green energy systems. The book embraces a broad variety of topics corresponds to the latest changes and emerging trends in the industry, including but not limited to:

- Chapter 1: Synthesis, electrical setup, sustainable design, as well as energy storage and production related to Carbon Nanomaterials in the framework of Green Energy systems are outlined;
- Chapter 2: The environmental impact of carbon nanostructures which are part and parcel of fuel cells is quantified giving useful information;
- Chapter 3: Advanced Nanomaterials in Lithium-ion Battery: The innovation in nanomaterials is steering significant progress in the storage capacity and electrochemical performance of Li-ion batteries;
- Chapter 4: Research on Nanoengineered Materials to use in Solar Cells: planning and creating different layouts of nano-enabled solar panels, evaluating their benefits, drawbacks, and possible improvements.

Besides main concepts and methods of synthesis, the use of nanoengineered materials in modern energy technologies will be underlined in this book. The problems that should be addressed are about the environment, the possibility of producing the materials in mass, and the duration of the materials. This research embodies the use of nanoengineering strategies to drive innovation with regard to the next generation energy technologies with the objective of working towards what can be described as energy that can be described as cost-effective, energy-efficient, and environmentally friendly-aligning with the agenda of green energy proliferation.

Assoc. Prof. Dr. Rashad Gabil Oglu Abaszade
Assoc. Prof. Dr. A. K. Haghi

Contents

About the Authors

Dr. Rashad Gabil Oglu Abaszade since 2022 serves as an Associate Professor at the University of Azerbaijan State Oil and Industry and since 2024 at Azerbaijan University of Architecture and Construction. He is a Physics Ph.D. (2018) who has already received the degrees in Physical Electronics (B.Sc., 2003) and Optoelectronics (M.Sc., 2006). During his academic career, he has published 153 research articles, nine textbooks as well as so many academic curriculums and has been awarded six patents. According to Google Scholar, his h-index is 18 and this author has been cited more than 933 times. Dr. Abaszade is the director of the Department of Innovation Research at International Academy of Ecoenergetics, where he conducts research on synthesis, characterization and application of nanomaterials, especially graphene-based systems, established in 2019. He has worked in such areas as development of nanoelectronics including elaboration of nano-components like negatrons, diodes, capacitors, and transistors making use of advanced carbon-based nanostructures. The nanomatrix devices are under continuous development in the research group of Abaszade, where smart materials are integrated in order to apply them in smart city infrastructure and the alternative energy stations. His initiatives are conducted together with the research centers in the USA, Germany, Belarus, Kazakhstan, Turkey, Pakistan, Poland, Portugal, Mexico, Uzbekistan, Russia, Ukraine, Hungary, the UK, Chile, China, India, and Italy. He holds the position of Editor-in-Chief at Ecoenergetics and regularly contributes as a peer reviewer and editorial advisor for numerous high-ranking international journals. In 2019, he received Professor title of the International Academy of Ecoenergetics, and since 2023 he has also been a visiting professor of Nursultan Nazarbayev University. Dr. Rashad Gabil Oglu Abaszade is also the director of the Turan International Research Institute.

Dr. A. K. Haghi One of the respected scientists with very high level of academic output (more than 1000 scientific works were produced under category of authorship or coauthorship, reviewing or coediting, or books and chapters), As it is described by google scholar, his work has resulted in over 4100 citations and his h-index is 34 that describes his contribution to ISP aspects in different areas of science. Dr. Haghi occupies honorary academic professorships in some well-known institutions. At the international level he has collaborations in Mexico and Portugal as Honorary Research Associate and a Senior Research Fellowship in the Tbilisi State University, Georgia. He has directed, and supervised (acted as a mentor) a number of Ph.D. and M.Sc. theses during his career, and worked in joint international Ph.D. programmes. He alternatively played a very important role by directing postgraduates in the University of Guilan and also launching the exchange programs through academics. Dr. Haghi was an undergraduate in Urban and Environmental Engineer in the University of North Carolina (USA). He then received graduate degrees in the USA and France, succeeding a Master of Mechanical Engineering degree at North Carolina State University, as well as a Master of Applied Mechanics, Acoustics, and Materials degree at the Universite de Technologie de Compiegne. The most relevant of them is a Ph.D. in Engineering Sciences, which he received in 2001 by the Universite de Franche-Comte (France). Dr. Haghi is known by many to be an active speaker, keynote lecturer and participant in many scientific national and international conferences. Through his editorial activities he has founded and is Editor-in-Chief of the International Journal of Cheminformatics and Chemical Engineering as well as the Polymers Research Journal, both of which are in line with his past and present belief in the importance of interdisciplinary research and communication. He is also a peer reviewer in the top scientific journals of the world besides the fact that he is the editorial board member of most of the highest impact journals.

List of Figures

List of Tables

Classification of Carbon Nanotubes 1

Structurally, one can consider carbon nanotubes as sheets of graphene that are rolled into cylinders without any disruptions and carbon atoms arranged in hexagonal patterns. Their mechanical, electrical and structural characteristics have made them some of the significant materials in most of the applications of nanotechnology and electronics products. Among the most important issues of the way of cognition of CNTs is its classification that is based primarily on the number of layers of graphene and geometric topology of carbon atoms within a cylindrical nanostructure. The precise nature of structural classification of carbon nanotube relies primarily on the number of the graphene sheets which are provoked into formality where they are introduced as cylinder (Table 1.1).

Chirality is widely occurring in nature and has been seen to play major roles in biological and materials sciences. Chiral recognition defines the non-identical response of the chiral guest with the chiral host. In the specified case, one can point out a wide application of nanomaterials in chiral sphere. Carbon nanoparticle is the most utilized and researched of all. The mostly utilized carbon nanomaterials are mostly grapheme, graphene quantum dots fullerenes and carbon nanotubes. Chirality is mainly due to material carbon or chiral post modified molecules which obtain the carbon nanomaterials. When chirality and carbon nanomaterials are employed together the outcomes that are obtained are a material that has a chiral center as well as an improved conductivity and a propensity to biocompatibility. What is implied here is that the produced synthesis of the chiral carbon compounds can be applied to the greatest extent in many spheres, and these spheres are the sphere of chemistry, biology, medicine and pharmacy. The paper is addressing the current progress of the chiral carbon materials following a number of the types of carbon nanomaterials, synthesis and origin of the chirality of the carbon-basis chiral materials. Its use in sensors, disease therapy, asymmetric catalysis, chromatographic chiral separations

R. G. O. Abaszade and A. K. Haghi, *Carbon Nanotube Devices for Nanoelectronics*,
Synthesis Lectures on Solid State Materials and Devices,
https://doi.org/10.1007/978-3-032-02114-4_1

Table 1.1 Classification of carbon nanotubes and their general characteristics

Type	Structure	Diameter	Length	Electronic properties	Properties	Application
Single-walled carbon nanotubes (SWCNTs)	Single cylindrical graphene layer	~ 0.4–2 nm	Several micrometers	Metallic or semiconducting (chirality-dependent)	Large flexibility, excellent electronics (electrical) qualities, flexible electronic properties (metallic/ semiconducting)	Nanoelectronics, sensors, batteries
Double-walled CNTs (DWCNTs) (special class of MWCNTs)	Two concentric graphene cylinders	~ 1.5–5 nm	Several micrometers	Metallic or semiconducting	Bandstructure intermediary properties, strengthened conducting mechanical strength but more control on the electronic properties	Biomedical, advanced composites
Multi-walled carbon Nanotubes (MWCNTs)	Multiple concentric graphene cylinders	~ 2–100 nm	Tens to hundreds of micrometers	Predominantly metallic	Excellence in mechanical strength, thermal conductivity and excellent chemical stability	Structural composites, energy storage

are also described. Lastly, we give some reflections on challenges and opportunities that exist in front of chiral carbon materials.

CNTs can be positioned in a geometric context in addition to the sorting of the number of walls [1–3]. The involved gas sensing method will have a possibility to develop far-smarter, and more preferential miniature sensing systems without necessitating any special materials in the course of coating and functionalization. The electrical properties of carbon nanotubes to a great extent depend on the chiral angle and the symmetry of layers of graphene folded [4, 5]. Since the emergence of interest in the biological utilization of nanoparticles (NPs), more concerns on how safe and toxic they are having been raised. Nanoparticle toxicity and how it behaves in the biological system also matters and it is important to have a clear picture about the nanoparticle behavior and toxicity in order to arrive at safer and effective nanomaterials. Further, another use of nanoparticle as a method of imaging, delivering the gene, tissue engineering, targeted therapy and vaccine development are mentioned in this review. The manner in which nanoparticle (NP) toxicity is attained is varied and there are many factors that condition toxicity and behavior of NPs and they have been rigorously outlined in this article. Such studies gaze at the influence of physicochemical properties of nanoparticles (e.g., particle size, morphology, structural arrangement, aggregation, surface charge, hydrophilicity, concentration, material). Specifically, the toxicity of the various types of nanoparticles, e.g., polymeric, silica-based, carbon-based, metallic, and plasmonic alloy nanoparticles, was studied in a systematic way. In the classification of carbon nanotubes, one also needs to research their interactions in the environment and with the biological systems. There is a possibility that carbon nanotubes (CNTs) toxicity and compatibility may depend on their surface cover and form, because they are used in medic and environmental systems. Melanoma is a skin carcinoma (cancerous) and it occurs in over 140 000 patients annually across the globe with the death rate being high. In the worse instances, treatment remedies are ineffective. It is therefore most essential and imperative to find alternative forms of treatment, i.e. immunotherapies, to treat, alleviate and/or prevent cancer. The conjugation of carbon nanoparticle to different products of plants like with aloe vera has shown good antineoplastic effect, probably as a result of the existence of chemicals a loin and others like aloe-emodin, barbaloin, acemannan, and octapeptide, therefore opening up potential of use with others as antitumor entities. This reading discusses the failures and advances of the treatment of melanoma with the functionalized carbon nanoparticles. The analysis gives a platform upon which further studies are required to conduct an in-depth analysis on the potentiality of carbon nanomaterials together with Aloe vera in treating different types of malignancies including melanoma [6, 7].

According to the structure, morphology, electrical properties and environmental impacts, there are many groups of carbon nanotubes. All the characteristics of CNTs and their fields of application are largely determined by the number of its walls. New opportunities and possibilities to apply the specific features of carbon nanotubes to the different spheres of scientific and industrial life appear because the systematic study of these materials may provide new opportunities to the scientists and engineers.

1.1 Introduction

They have said that carbon nanotube has received much fame since it was discovered in early nineties due to its outstanding engineering, electrical and thermal reliability. They fall under the category of sophisticated compositions in current nanotechnology, since they are composed of carbon atoms that have been rolled- up as hybridized sp^2 atoms in fines into tubular features of the nanometer size. Carbon nanotubes have an amazing strength in tension, good electric conductivity and promising peculiarities in the electronic characterization, electronics, material science and health. The categorization of CNT is very essential in deep understanding thereby essential in the optimization of their characteristics and improvement of their performance in so many implementations being done.

In this chapter an elaborate discussion between the different structures of CNTs and the distinction between the structures of CNTs and the properties is provided. The tensile strength of carbon nanotubes is nearly extraordinary, are electrically conductive and possess excellent electronics characteristics, thus rendering an incredibly high flexibility to a broad array of applications in the fields of energy storage and electronics, materials science and healthcare applications. Adequate understanding of the classification of CNT is quite essential in maximizing its properties and improvement of performance in the majority of applications. All the types of structures of CNTs and the difference between the structures and the properties are elaborated in the chapter.

The Aspect ratio of carbon nanotubes (length to diameter ratio) is also a classification parameter. carbon nanotubes (CNTs) are classified as short (length/diameter ratio $L/D < 10$), intermediate ($10 < L/D < 50$), and long ($L/D > 50$) and this directly affects their mechanical and electrical properties. Generally, long nanotubes are more conductive and stronger than the short ones. The mentioned models are working ones actually and are attained as an outcome of certain boundary conditions and a personal prescription. Some of the literature results are as well theoretically confirmed by us namely existence of Dirac cones in 6,6,12- and 6,6,12- graphene nets and also in 6,6,12- graphene nets. We too were able to find a band gap, as anticipated in literature, but only over a limited parameter range and there is a transition with quadratic touch, beyond which we find Dirac cones occurring with 2-d 2-spin 2-orbital graphene using 3-d 2-spin 2-orbital we believe the case must be the same. Senthilkumar [8]. Their production methods of CNTs are classified based on the synthesis processes as well as the properties that are specific to the CNTs. The two protocols are capable of delivering multiple carbon nanotubes in relation to morphology besides providing purity [8–10].

1.2 Different Types of CNTs

Carbon nanotubes were among the most researched nanomaterials within the last few years because of the exceptional structural, chemical, and conductive characteristics. All the forms possess characteristic features in terms of structure and functionality by which the areas of their application to the field of electronics, materials science, energy storage, and biology are defined. Carbon nanotubes are typically characterized by the number of concentric sheets of graphene, stacked one above the other, in tube shape. Carbon nanotubes may either take the form of single-walled, multi-walled, or multi-walled depending on wall structure which possesses different physical and electronic properties. The fundamental classification can be further divided into different categories of chirality, diameter and length- the parameters that are vital in influencing the physical and chemical characteristics of the carbon nanotubes. Basic classes of CNTs and their peculiar features are depicted in the table below: In conclusion, modified doped type of normal copper and tungsten alloy material can be prepared, graphene produced in most situations and the properties of the material achieved in electrical conductivity and hardness advanced and bandwidth in the design and production of high voltage type of electrical connections extended exponentially. The addition of graphene to the mixture also has the effect of increasing electrical characteristics of the copper tungsten electric contacts as well as that of corrosion resistance; therefore, the current study provides a theoretical deduction and methodological advice in relation to the choice as well as fabrication of materials employed in high voltage, high current circuit breakers. The innovation contributes immensely to the graphene technology and high-voltage electrical contact technology. They have either a metallic or semiconducting behavior and this aspect is provided by its peculiar structure mainly caused by the chirality concerning the geometry of the location of carbon atoms with which the tube is made. It is possible to describe chirality by the pair of integers (n, m) after which the graphene sheet should be rolled into a vain nanotube along its lattice vectors in how many ways and to what extent. In this paper, the history of LiO_2 battery will be traced and its introduction will be mentioned. The lithium-oxygen battery is one of the much-hyped high-performance batteries, and the solution to the energy storage capacity of batteries problem. Despite the fact that lithium-ion batteries are being still utilized, lithium-oxygen batteries are much more superior by the measure of their energy storage capability. The factor behind the working mechanism and the underlying design of lithium-oxygen battery is also highlighted. Additional details together with the solid electrolyte interphase and the lithium dendrites are also mentioned. Nevertheless, it has several architectures which continue to provide more performance options to make use of. This article creates a positive note on the future of development of LiO_2 batteries. Well, when $n = m$, carbon nanotubes behave like metals, and otherwise, they tend to have the semiconductor properties [11, 12].

The last few years have revealed that nanocomposites material can be feasible options in a number of applications in diverse industries due to their interesting characteristics.

This project aims at examining how some of the properties of polymers can be changed by adding carbon nanotubes, where the great surface area to volume ratio of the MWCNTs is used to improve their properties.

To this end, MWCNTs were incorporated in different proportions and with the help of different mixing methods. Mechanical, electrical, and thermal characterizations of the resulting nanocomposites materials were used to measure the effect of this addition. Homogeneous distribution of MWCNT boosts the general characteristics of the material so that it can productively make use of its special features. The results demonstrated that incorporating a low concentration of MWCNTs resulted in uniform distribution and dispersion within the matrix, which in turn enhanced key material properties, including tensile strength, elongation at break, and Young's modulus. A weight percentage ratio of 0.5 wt% of MWCNTs represents an average ratio of MWCNTs addition into different types of polymers. Such composite nanoparticles will be able to replace numerous industrial materials in a myriad of applications. Single-walled carbon nanotubes have sizes that often range between 0.7 nm to 2 nm in diameter, and could be several micrometers in length, which makes them have large aspect ratios in different applications [13] (Fig. 1.1).

MWCNTs however have several concentric layers of carbon nanotubes, and the thickness of the layer may vary in a few to dozens of layers and is bound by van der Waals force. Due to their multilayer structure, MWCNTs are Mechanically stronger and better thermally conducting than SWCNTs, but at the cost of general worse electronic behavior. In this paper, the author is going to inform about carbon nanoparticle as a potential

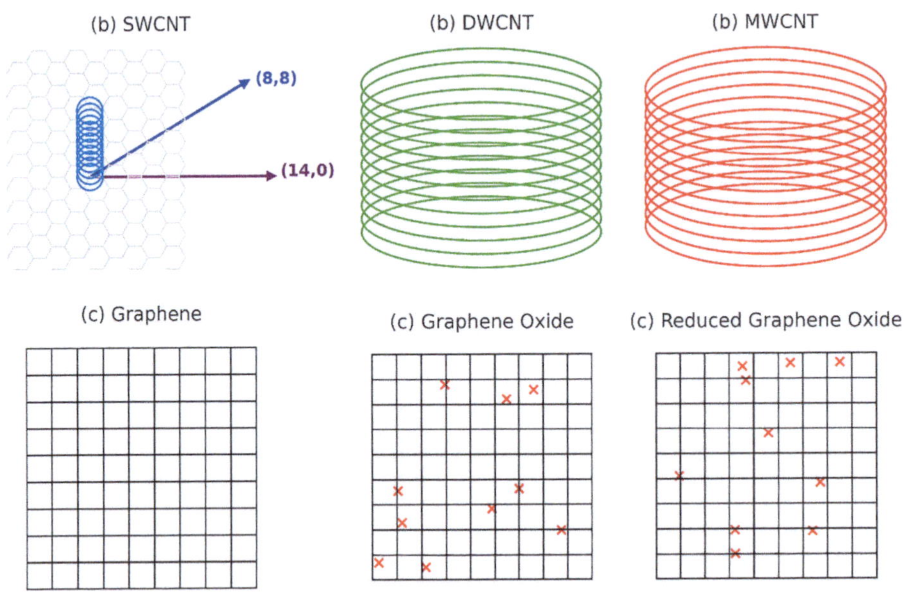

Fig. 1.1 Graphene-based materials

antibacterial and food package material. Petroleum based products, synthetic material, ceramics, wax and other materials required in food-packaging industry generate emission of polluted gases and waste water which result to contamination of environment. Compared to the current challenges facing the preservation and packaging of foods, carbon nanoparticles may simply provide equally excellent alternatives to them hence increasing the expiry dates of foods without affecting their nutritive values. The number of carbon nanomaterials (CNMs) is numerous because the carbon atoms are connected in different structures to form many allotropes. Carbon nanomaterials (CNMs) maybe, in general, categorized depending on their dimension type namely zero-dimensional that imparts carbon dots and graphene quantum dots, 0D, 1D, 2D, and others of graphene type derivatives.

They are also antimicrobial and that quality can be applied as food preservative and many others. Therefore, carbon nanostructures are highly useful. It is possible that a MWCNT is cheaper to produce than a SWCNT, a factor that is favorable in a number of applications such as composites and nanoelectronics [14, 15] (Fig. 1.2).

The difference between the SWCNTs and MWCNTs also includes the method of production. The self-assembly offers certain prospects in fabrication of nanostructures, of a morphology and property varied, and generally made with vapour phase or liquid phase as precursor material The new incarnations of the drug delivery system, of silicon nanoelectronics, laser, nanotechnology creations on the basis of nanoscale self-assembly, adaptive property, simple and scalable are all the more such that increasing presence and future potential of more who-sy-what nanostructures. The objective of the study is to give the reader a sketch pen picture of what nanoscale self-assembly is all about, the current state of application of the same and the future of application of the same. Particularly, the CVD method has been very effective in the instance of SWCNTs where aligned nanotubes of specific chirality in high purity have been grown [16].

Single-molecule magnets (SMMs)-functionalized carbon nanotubes (CNTs) have shown a potentiality of enhancing the molecular magnetism and will offer new magnetic materials. It is also through SMMs and CNTs that an extremely thrilling route towards the development and fabrication of new magnetic materials whose properties can be customized has been made possible. The view of such synergetic effects that take place under the influence of such hybridization is the perspective object of investigation in the fields of basic science of molecular magnetism and the development of the new application in the fields of nanotechnology. This type of hybrid systems enables a synergetic mixing of SMMs and CNTs which introduces a train of remarkable advantages. Superior kinetical properties of CNTs present the perfect material to encapsulate and pattern SMMs hence organize their assemblage and well-regulated actions. Moreover, hybridization generally guarantees new functionality and better performance, both because of interaction of magnetic moments of single-molecule magnets, and because of the electronic structure of carbon nanotubes. However, Challenges and opportunities regardless of the fact that there has already been made a tremendous progress in direction of the synthesis, characterization and applications of SMM-CNT hybrids, yet there are some. Experiment on

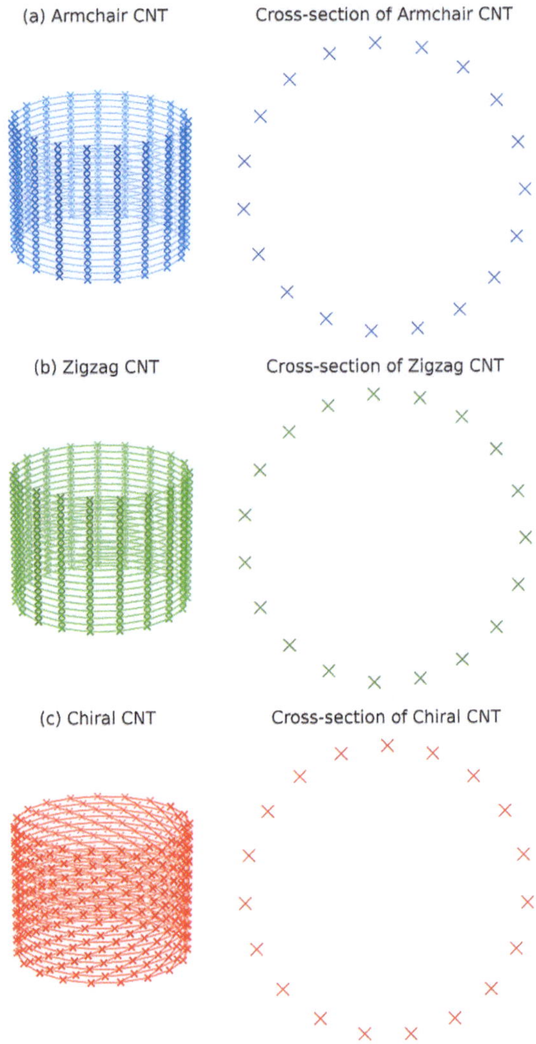

Fig. 1.2 Models of nanotubes: armchair (**a**), zigzag (**b**), and chiral (**c**)

this hand still needs to be carried out to extend what is already known about the basic interaction between SMMs and CNTs, as well as to be able to design and synthesize such hybrid systems at some later time, and to discover other means of controlling their characteristics. Additional criteria toward no only the progress of the fabrication procedure but also toward the methodology of the characterization will enable us to engineer more complicated hybrid constructions and will induce the further understanding of the relationship between the structure and the property. Growing MWCNTs may be produced under much more wide-ranging settings, which may imply less expertise over chirality, however in many cases mighty better productivity [17].

Table 1.2 Classification of carbon nanotubes according to chiral vector (n, m) values and corresponding chirality angles

Type	(n,m) Condition	Chirality angle θ
Armchair	n = m	30°
Zigzag	m = 0 or n = 0	0°
Chiral	n ≠ m, n ≠ 0, m ≠ 0	0° < θ < 30°

The two structural templates that CNTs are broadly classified are the SWCNs the and MWCNs. Even more detail is known as the scheme of classification taking into account also the chirality (roll structure of the sheet of graphene) the diameter and the number of walls. An important criterion of determining carbon nanotubes (CNTs) is the chiral vector that defines the wrapping structure of the graphene sheet.

$$\vec{C_h} = n\vec{a_1} + m\vec{a_2}$$

where:

$\vec{C_h}$—chiral vector;

n, m—integest defining the tube structure;

$\vec{a_1}, \vec{a_2}$—unit vector of the graphene lattice.

From (n, m), the chiral angle θ and tube diameter d_1 can be calculated:

$$\theta = tan^{-1}\left(\frac{\sqrt{3}m}{2n + m}\right)$$

$$d_1 = \frac{\alpha}{\pi}\sqrt{n^2 + m^2 + nm}$$

where:

$a = 0,246\,nm$—lattice constant of graphene.

Classification based on (n, m), (Table 1.2):

In addition to this, geometry and surface chemistry of CNTs is unique and could be modified using different approaches to synthesize and functionalization techniques. A catalyst synthesis procedure using a combination of the precipitation and impregnation procedure was realized using modified Stober based approach using a gel. The yields of biodiesel above 80% were achieved in the transesterification process in which 3 h were utilized. The existence of a possibility to modify CNTs and, consequently, improve them makes their potential in terms of application in more advanced applications practically unlimited, including the possibility to develop lightweight composite materials with mechanical characteristics that would be superior to them and nanowing structures that will serve as the building blocks of electronic devices whose properties will be at the nanometer level [18].

In electronics, carbon nanotubes (CNTs) feature in the development of field-effect transistors (CNTFETs), owing to the good conductivity and high mobility of the electrons they enjoy. It has found wide application in electronic, health as well as environmental studies, due to its enhanced mechanical properties. This has caused the search in the production of carbon nanotube transistor and its application in integrated circuit to be among the greatest theme of research over the last few years. In this article we are verifying the structure, production philosophy as well as several matters that are connected with the carbon nanotube field- effect transistors. It was stated how incineration and elimination of metallic carbon nanotubes can be conducted, since such contaminants are sure to be produced during the manufacturing process. This is then followed by the paper suggesting three methods of fabricating conductive channels. The final paper session discussed electrical properties of carbon nanotube FET. The change in the electrical properties under two types of deformations i.e., the axial as well as the rotational one is also analyzed [19]. Optical peculiarities Unique optical properties of CNTs have also found their way to be exploited in photodetectors and solar cells where they enhance the efficiency of charge transfer and energy conversion processes. The electric properties of carbon nanotubes (CNTs) are superior. The question, however, is with the full expression of these aspects in realistic electrochemical experiments. Prior art has improved electrical performances of carbon nanotube through welding. The conductivity boost mechanism, though, is obscure. Welding is a process that lacks thorough mechanics investigations and theory. The study brings out a theoretical layout of a Circuit made of carbon nanotubes with staggered electrodes and studies the impact on the twist angle to a bundle of carbon nanotubes. A CNT bundle circuit model welding is developed and described through the ratio of resistances of the CNT circuits prior along with post weld, in view of imagined architectural variations of CNT more than the whole course of action. An electrical characterization tool of the CNTs was to demonstrate the model of welding with the nanomanipulation apparatus that was applicable to a scanning electron microscope. The long electric welding experiment was performed and it was noticed that the conductivity raised in about 1.5 to 4 times the value after welding [20].

Of concern due to long-term perspective is the sustainability of the world supply of water, where the most significant variables are purity and safety. To curb this looming problem, adsorption technique has become one of the biggest and widely accepted Adsorption method or tools of choice regarding removal of hazardous chemicals and more specifically the lead removal in the waste treated water. This critical review deals with the current condition of art in the field of adsorption, in which much attention is paid to the radical stage of development of the field of separation processes and purification that is not confined to the common oxides of metals adsorbents. There are high chances that these materials may change their structures, at the nano level and the impressive finding is how fast they appear to remove the pollutants cleaning the water in the shortest time. The literature search of the work was done in the Google Scholar database, and the phase adsorbents lead remediation was used that brought approximately 6000 publications. It was also reduced to the last three years and then only review article which were pertaining to lead remediation. Over 100 articles were consulted to learn how to approach different methodologies, surface modification, and absorbent material in the remediation of water contamination by inorganic contaminants. In this experiment, this has shown that the study has its limitation especially when using the starch-based adsorbents in lead remediation. Turning to the fact of commercial exploitation the following review presents the problems that have been and are being met in the work of preparing inorganic adsorbents and points out in a critical way the prospects of the future. The promising direction is the surface modification and adsorption capacity may be increased significantly, perhaps, twice or even four times. Apart to this the adsorbents also have an amazing regeneration forces whereby even after a series of cycles a maximum of 90% of the regeneration power is still available. The starch adsorbents certainly have a great potential that should be worthwhile to be effective during the removal of lead in the water systems. Nevertheless, there is yet more research still needed, and this should be aimed at optimizing adsorption process and stability of the starch-based adsorbents under the realistic conditions. The slowness of the formation kinetics refers to the restrictions on usage of hydrates in practical purposes, which, together with the low salt amount of storage does not ensure the use of hydrates in industry, in processes of gas separation and transportation. In the present work the kinetic of CO_2 separation and hydrate formation in $CH_4 + CO_2$ (70:30) mixture and the influence of added MWCNT, OH-MWCNT, and COOH-MWCNT is investigated experimentally. The nanomaterials have their percentage weight maintained at 0.01–0.05 in solutions of 0.03 weight percent of SDS, or the ultra-pure water. It was detected that carbon nanotubes have a slight effect of increasing the induction time and the initial rate of gas consumption at the pressure 4.0 MPa and temperature 274.15 K. A successful CO_2 separation has been obtained and it was actually discovered that the separation factor was between 2.9 to 3.4 with the CO_2 recovery factor of about 65% on the $CH_4 + CO_2$(70:30) gaseous mixture. Between the rest of the samples, COOH-MWCNT had an increased effectiveness of the separation. Notably, COOH-MWCNT nanofluid, lacking the presence of surfactants, represented a significant level of selectivity towards CO_2 in contrast with a solution possessing 0.03 wt% surfactant ratio as determined [21, 22] (Fig. 1.3).

Besides that, CNTs were found to possess promising prospects in biomedical application, especially drug delivery and biosensors, which can be modified to be more bioactive and biocompatible. High porosity, biodegradability, structural stability, non-toxicity, and many other natural advantages of chitosan are denoted in most articles. Nanoparticle functionalization of chitosan has also been observed to enhance its characteristics as well its applicability in several applications. Composite materials made by chitosan, more particularly chitosan-based polymer nanocomposites, have drawn huge consideration and could be employed as superb adsorbents in dyes, oil spills and heavy metal ions, including cobalt, mercury and copper. What is more, these composite materials composed of chitosan exhibit the exceptional capability to be multifunctional due to the synergy effect of its components. The efficiency of adsorption has also been proved to be enhanced by the electron rich functional groups on the polymer backbone. In general, the properties

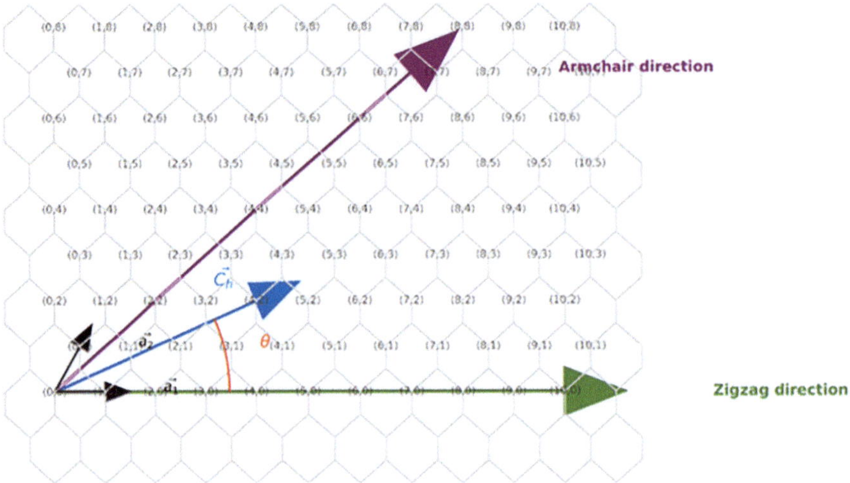

Fig. 1.3 A two-dimensional graphene layer to explain the chiral vector $\overrightarrow{C_h}$ and the chiral angle θ

of these chitosan-based nanocomposites synthesized by functionalization of chitosan with carbon nanostructures (carbon nanotubes) and nanoparticles (Ag, TiO_2, ZnO, and S) are amazing, and they can be the candidate materials that can be applied in the environmental and wastewater remediation [23].

Such capability to functionalize the characteristics of CNTs progresses the application of nanomedicine, which remains to be critically important to enhance diagnostic and treatment procedures. Concluding, there are two main categories of CNTs: SWCNTs, and MWCNTs and have critical differences in the structure, electrical, and mechanical characteristics hence endowing them with different uses. As the research on production and functionalization of the carbon nanotubes continues, it is expected that their use in technology and medicine will expand further and provide solutions in a number of fields.

1.2.1 Introduction to SWCNTs

While numerous layers of graphene brought together can undo a cylinder to be rigid, the one layer of graphene included in SWCNTs is only rolled into a cylinder. The SWCNTs are possible to be achieved due to the fact that a sheet of graphene can be rolled to form a single continuous cylinder. They are likely to be a few nanometers to some micrometers in length, their diameter being typically of the order of 0.4 nm to 2 nm. The extraordinary characteristic of the SWCNTs is the chirality that classifies it as a metallic or semi-conducting tube. SWCNTs are potential nanoelectronics, high-performance transistor and

Table 1.3 Electrical conductivity of CNT–polymer nanocomposites at different filler loadings

Property	Description
Structure	Single cylindrical graphene sheet
Diameter	~0.4–2 nm
Length	Typically, several micrometers
Chirality	Determines electronic properties (metallic/semiconducting)
Electrical conductivity	High; tunable through chirality and doping
Mechanical Strength	High tensile strength and flexibility
Thermal conductivity	Very high along the tube axis (~3500 W/m·K)
Typical applications	Nanoelectronics, transistors, sensors, drug delivery

biosensor materials due to extremely tunable electrical characteristics and low defects levels. In addition to that they are elastic and possess a large area of coverage hence making them of interest to be used in energy storage and delivery of drugs (Table 1.3).

The carbon nanotubes (SWCNTs) are the fundamental nanomaterials of the modern era of materials research with unique combination of mechano-electrical-thermal properties. Their geometry is odd, a one-dimensional rolled-up sheet of graphene, and so they exhibit behaviors caused by quantum confinement effects and by chirality-dependent properties. The explicit knowledge of how SWCNTs are produced, what their properties are, and how they can be utilized gives an essential insight into the information concerning their utility in the technologies of the future, such as flexible electronic devices and the next-generation energy storage tools. Their mechanical properties which provide them with outstanding values of conductivity and strength make SWCNTs very appealing in research (academic and industry). Their normal structure, i.e. chirality and diameter, has a profound effect on the properties of SWCNTs and determine their electrically conductive nature; SWCNTs may be metallic or semiconducting, according to their chirality and diameter. The special case where (n - m = 3l), and (l) is an integer, produces a single-walled carbon nanotube (SWCNT) that is quasi-metallic; otherwise the property is semiconducting. The energy dispersion characteristics of SWCNT films are influenced by their longitudinal quasi-momentum (q), nanotube density, and film thickness. Modifying the thickness allows for tunability of exciton energy levels within the film. Additionally, the excitonic band gap—which reflects the transition energy—varies depending on incident photon momentum, dielectric properties of the surrounding media (including substrate and superstrate), and film thickness. At T = 0 the excitonic gap of a thin sheet of periodically arranged (11,0) SWCNTs should be around 1.116 eV [24].

The numerous electronics characteristics of SWCNTs take on its potential suitability in diverse ways, including transistors and sensors, in which the conductivity qualities of the material may be tuned. The manufacturing has its processes so vital in the quality of SWCNTs and its capability. The most widely used SWCNT synthesis technique is

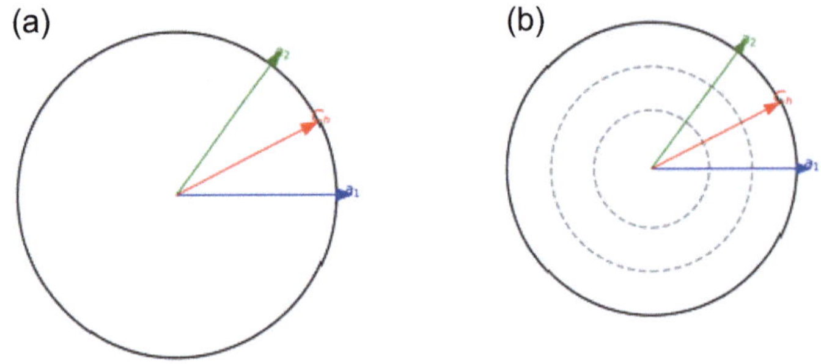

Fig. 1.4 Chirality of SWCNTs (**a**) MWCNTs (**b**)

the CVD approach whereby, metal-based catalysts thermally decompose hydrocarbons allowing the nanotube to grow under very restricted conditions. The synthesis experiment involved the insertion of the ZIF-67 dispersion in ethanol-based solution in a tube furnace at 1100 °C with a hydrogen ambiance. ZIF-67 was used as a catalyst precursor, ethanol was the carbon source, and H_2 was the carrier gas. The presence of H_2 around ZIF-67 inhibits the rapid aggregation of the Co nano catalysts, however ethanol considerably facilitates the decomposition of ZIF-67 and the growth of SWCNTs during FCCVD. The SWCNTs synthesized are of monodisperse diameter and high purity that can be used to construct various macrostructures as well as an effective photocatalyst in antibacterial electrocatalysis. Using ethanol in allowance of the traditionally harmful gaseous carbon precursors would not only make the synthesis process cheaper but cost-effective, this would enable the process to be used as an environmentally friendly method of producing nanomaterials on a larger scale [25] (Fig. 1.4).

Fine tuning of this synthesis can produce high aspect ratio SWCNTs of controlled dimensions and purification can be used to enlarge the yield of pure structures. The structurally possible single-wall carbon nanotubes with the smallest diameters are desired in the challenges that require electrical band gap greater than 1 eV. To investigate the influence of ruthenium promotion on cobalt-based catalysts in the synthesis of small-diameter single-walled carbon nanotubes, we employed a high-throughput, fully automated laser-assisted chemical vapor deposition (CVD) platform integrated with real-time Raman spectroscopy monitoring—referred to as the Autonomous Research System (ARES). By comparing more than 200 growth experiments by Co and the Ru-promoted Co catalyst at 750 °C over a wide range of feedstocks to multi-excitation Raman analysis of their products, we have demonstrated that the selectivity to small-diameter (diameters < 1 nm) SWCNTs is approximately doubled by the Ru-promoted Co catalyst. Ruthenium is a high temperature cured (800–850 °C) cobalt catalyst nanoparticle which increases the selectivity of small diameter single-wall carbon nanotubes by approximately a factor. Commutes of Co_xRu_y clusters placed on 13 and 55 atoms were evaluated by using density functional theory (DFT) and the range of Ru content was zero to 22%. The obtained results demonstrate that the stronger the concentration of Ru is, the more the cohesive energy is, irrespective of the position of the Ru- atoms in the cluster. Their results indicated that Ru content enhanced the melting temperature and decreased the atom mobility in agreement with the Ru content of about 10% in our Co catalyst that would raise the obstacle to sintering and stabilize small nanoparticles and could be the source of the high selectivity to small-diameter SWCNTs produced. Second, the simultaneous development of various catalysts might allow the selective development of semiconducting or metallic tubes, which would be necessary to connect to electronics [26].

The process of functionalization increases the solubility, dispersibility, chemical reactivity as well as the compatibility of CNTs in various environments or matrices. The diagram focuses in the pristine carbon nanotube at the middle, which may undergo a variety of functionalization. The approaches are mainly classified in the following routes (Fig. 1.5):

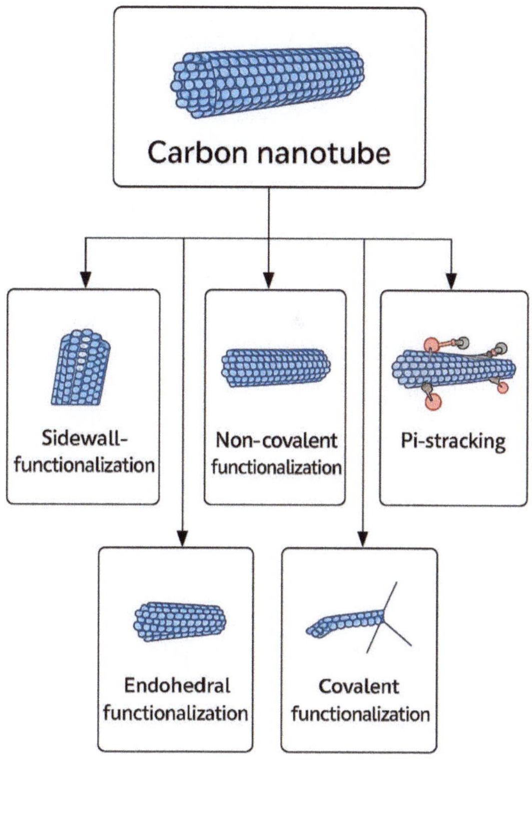

Fig. 1.5 Classification of functionalization strategies for carbon nanotubes

- *Sidewall Functionalization: In this technique, external cylindrical surface of the CNTs undergoes chemical modification. Side wall functionalization commonly takes place in oxidation cleavage, fluorination or even cycle additions.*
- *Non-Covalent Functionalization: Molecules are attached to the CNT surface by relatively (weak) non-covalent interactions, either by 1) 11/12/13. The native electronic properties of CNTs are maintained in this method that is largely applied in biosensing and biofunctionalization. The common agents are polymers, surfactant and aromatic molecules.*
- *π–π Stacking: Pi-pi stacking as a special form of non-covalent binding, it is the interaction between clouds of pi electrons of CNTs and aromatic structures (e.g. benzene rings). It enables functional molecules to attach to the surface of nanotubes thus leaving the pattern of the naphthalene sp^2-hybridized arrangement of carbon intact. This is beneficial to those applications that need the maintenance of electrical conductivity.*
- *Endohedral Functionalization: It entails containerization of molecules, atoms, or ions within the empty space at the interior of the nanotube. Endohedral functionalization can be used to allow release of drugs, transport of molecules or nanoreactor systems. It is arrived through opening the tips of the CNTs or capillary fill technique.*
- *Covalent Functionalization: Under this process, functional groups are involved in the formation of powerful covalent bonds with the carbon atoms that make the CNT lattice through the defect sites or endings. Indeed, although this may break the conjugated system of the CNTs, this leads to the strong chemical bonding, so is the perfect candidate as structural composites or as chemical sensors. A typical reaction to them is amidation, esterification, and radical reaction.*

The schematic is a systematic image showing major functionalization strategies that can be adopted in the use of carbon nanotubes. Method selection is based on the desired application and its necessity to preserve the original electronic structure of the nanotube or not. Strong binding of covalent methods result in altered conductivity and weak binding of non-covalent methods maintains conductivity. It is critical to comprehend and determine the right path to the engineering of CNT-based systems delivering the specified performance properties.

1.2.2 Functionalization of Carbon Nanotubes

Carbon nanotubes were first functionalized. 0.5g of high purity carbon nanotubes were obtained and purchased using the arc discharge method and then added into 250ml of 8M sulphuric acid into a 500mm beaker. The carbon nanotubes and sulphuric acid were continuously stirred in a heater magnet mixer. The reaction was carried out at 50 °C in 3 h. With the assistance of distilled water, the produced chemical (functionalized carbon

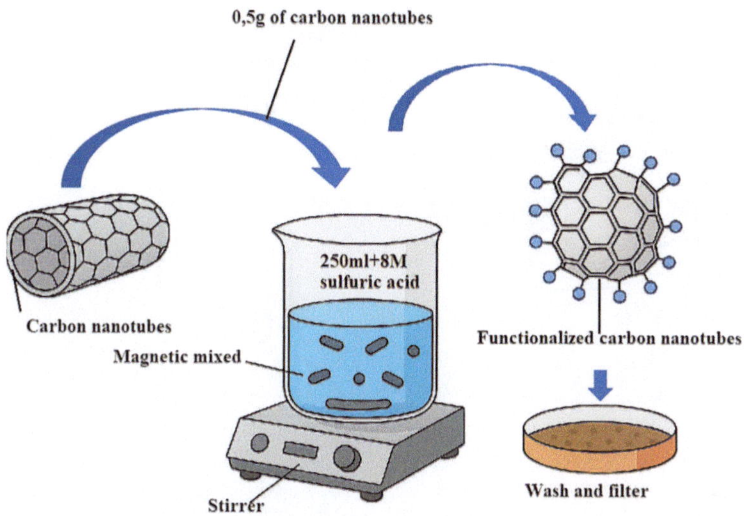

Fig. 1.6 Functionalized of carbon nanotubes

nanotubes) was washed and filtrated to obtain a pH of 7 then dried in a vacuum oven in a Petri dish [1, 2] (Fig. 1.6).

1.2.3 Doping with Gadolinium Process

The quantity of Gd_2S_3 used was 0.01 grammes and the quantity of the functionalized carbon nanotubes was 0.1 grammes (1–10%) mixed in a Petri plate and transferred into a 150-ml flask. This solution contained 50ml of distilled water. The mixture was constantly stirred in a magnetic machine which fitted with reflux condenser and a thermometer and left at 75 °C constant temperature of six hours. The new chemical was filtered, extracted and hoover dried [1, 2] (Fig. 1.7).

A synthetic graphite source material, graphene oxide large-scale, transparent, thin flakes have been synthesized by a modification of the Hummer 3 method. The SEM image analysis depicts that the thickness and the lateral size of the synthesized flakes of graphene oxide are 0. 31 10^{-3} m and 14.8 10^{-3} m, respectively. A decreased amount of the defective GO flakes was acquired, as the Raman analysis results indicate, and such flakes are greatly soluble in DMF, which is an extremely vital attribute of such flakes in terms of nanocomposites production. The EDX spectrum displayed the peaks of carbon and oxygen ingredients and the existence of oxygen confirmed that the graphene oxide was produced, and the graphene layers are functionalized with oxygen-containing groups. Their properties can make the GO flakes applicable in thin film and network, and

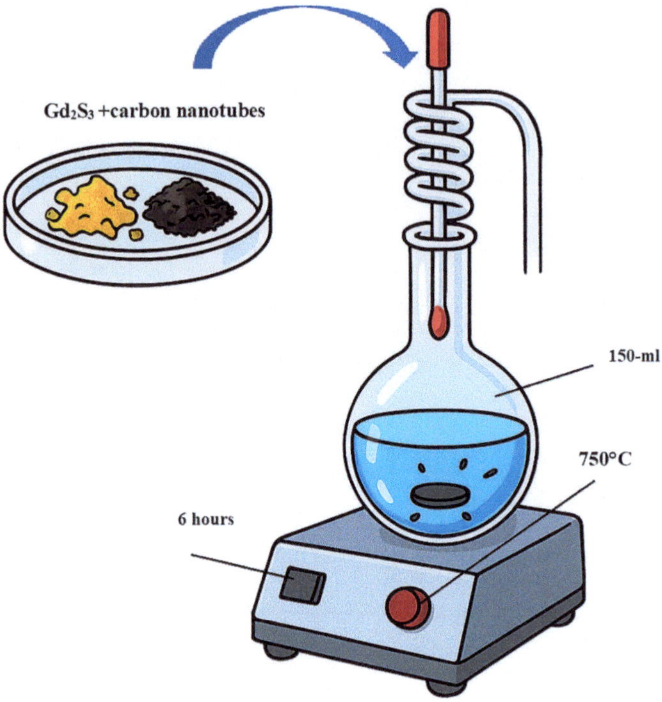

Fig. 1.7 Doping process in carbon nanotubes

the potential application in the transparent nanocomposite's technology, biomedical, and electro-magnetic interference environmental shielding microwave absorbers [27].

The synthesis can be noted as proceeding as follows: A beaker is added 10 ml of 96% concentrated sulphuric acid incrementally. 0.5g of NaOH is added to a petri dish well then 1g high-dispersion graphite at 99.9995% purity is mechanically added over solid sulphuric acid. The other chemical is added to the system in the amount of 13 grammes with the help of a magnetic stirrer after that solid sulphuric acid is added. The system is added in parts during a period of 2 h with potassium permanganate (3g). At this point the system is further placed in an ice bath and the temperature set to 20 °C. After cancellation of the stirring, the place is subjected to steam of a hot water bath, where the temperature is set to 35 °C. This would cause the mixture to change to the form of chestnut-grey paste. To it 46ml of distilled water is added. Mix 250ml distilled water and heat to boiling point in 109 boiling water baths. then after one leaf it sometime 100ml of perhydro is added and stirred. Thus, the visual observation of synthesis of graphene oxide is the release of yellowish substance in chestnuts. The mixture is filtered on cooling. The residue on filter paper is cleaned with distilled water. The product which was isolated at 50-degree temperature and 2 h vacuity was dried in a vacuum dryer during the process. The completed drying produces the product as a powder and the result is a 1.8-g product which is the overall mass loss of about 40% at the end of the process. X-Ray phase analysis was used to determine the structural analysis of the low-oxidized graphene oxide that was synthesized through the Hummers procedure. An XRD experiment was conducted by using D2 Phaser diffractometer (Bruker) with CuK_{alpha} radiation (l = 1.5406) between range of 2 variables 0.5° to 80°. This describes the graph of XRD pattern of GO sample. The 002 reflection in the flakes of graphite is sharp at −26.3 0 which marks its intermediary layer. Now let us suppose that we apply the term Debye Scherer to this sharp peak then the thickness of the graphite flakes would have been approx. 12.9nm thick and the number of layers would have been present here as compared to the number of layers we would have got. An Nd: YAG laser, 532 nm, and 10 mW max was used as a source of excitation. spectral resolution here is 0.5 sm^{-1}. A CCD camera was used as a detector, cooled to (−700 S) and used in a phonon counting mode. The Raman spectroscopy is a branched molecular spectroscopy of Raman scattered light which is an elastically scattered light of interest in the observation of phonons determination. The first peak in the spectrum is the D peak which is explained by the defects of the sample. The second peak is G peak. this peak is ascribed to the light scattering by Brillion zone. This peak includes doubly sheared peak as is usual in all specimens that possess sp^2 hybridization. Raman scattering curve shows that a sharp 2D peak appears when a layer of graphene exists, it is concluded that the graphene oxide is discrete with the thickness of different layers. Hence the D peak Raman analysis is 1320 cm^{-1}, the G peak is 1582 cm^{-1}, the 2D peak is 2720 cm^{-1}, the D + D/ peak is 2910 cm^{-1}, the 2D/ peak is 3190 cm^{-1}, the ID/IG ~ is 0.83 and the I2D/IG ~ is 0.496 respectively. In this case the 2D peak identifies the number of layers of the graphene. The result of Raman analysis shows that it obtained a high quality low-graphene oxide product. These Raman analyses findings are in accordance with the X-ray phase analysis findings. The values of the resistance are then noted down against change in temperature. The analysis of heating and cooling curve showed that the temperature was 9.84–6.5109 Ohm and 9.84–6.5109 Ohm respectively as the graph suggests. The cooling graph of the sample of graphene oxide shows an approximate stability of 10–25 °C. According to law of activation there is a sharp decrease in the range of 120–178 °C [28].

We have already developed a new method of synthesizing fewer and more functionalized layers of graphene oxide by Hammer method. The amount of graphite was 2 g, sodium nitrate ($NaNO_3$) was 1 g and potassium permanganate ($KMnO_4$) was 6 g, which was weighed using 3 separate plastic containers. Furthermore, a 46 ml of concentrated sulfuric acid (H_2SO4, 95 98) was measured with the help of a 250 ml graduated cylinder. The purpose as to why the ice was put in the ice bath was because of testing. The chosen methodology was utilized twice, since in the proposed chemistry two reactions take place. Flasks were naked in ice bath. $NaNO_3$ and H_2SO_4 as well as the weighted graphite were added in turn. The agitation is initiated and a thermometer is added into the flask. It still had 20 min to reach the temperature of 0 °C. The ice bath was removed at 0 °C. and this resulted in the increase in temperature of the mixture. This was heated in a hot plate at a higher rate of up to 20 °C. When the temperature reached 20 °C, KMnO4 was added slowly over two hours. This was maintained at 20–25 °C in an ice bath since it rose when $KMnO_4$ was added. After four more hours of adding all $KMnO_4$, the mixture was stirred once more. In this case the temperature was maintained at 20–25 °C. The mixture was then stirred at 35 °C upon reaching this temperature and this was then allowed to stay at 35 °C and the agitation was continued after 30 min. The amalgamation was put in ice bath. Deionized water (92 ml) was added to the sample in increments. In it the temperature was quickly brought to 90^0 C and then cooled to 70 °C. In this case we made 500 ml of 3% H_2O_2 and we added it to the mixture and we shook it over 10 min. The mixtures were settled 12 h, filtration is commenced. DW was applied by us in the filtration phase to eliminate sulphuric acid to clean our products [29].

One of the well-researched mechanisms of establishing and analyzing the chemical composition of a compound is the infrared spectroscopy. It may as well be used in identification of the functional groups present in graphene oxide. Infrared absorption spectroscopy was used to study the carbon nanotube samples in the wavenumber range of 4000 to 400 cm^{-1} based on a Varian 640-IR spectrometer. Graphene oxide can be clustered into some oxygen functional groups which include; alcohol, epoxy, carboxyl and carbonyl groups in Lerf-Klinowski model. The flakes of the graphene oxide show luminescence spectrum which suggests the presence of the following functional groups in the flakes: carbon-carbon bonds, carboxyl groups, carbonyl groups, alcohols, aromatic esters and epoxy groups. The spectrum 2007–2363 cm^{-1} contains mainly bands attributed to carboxyl and alkane and carbon–carbon functional groups. Their maximum at 1524, 1525, 1526 and 1558 cm^{-1} could be attributed to the benzene rings. The existence of the aromatic ether and alcohol functional groups is backed by the peaks at the area of approximately 1003–1063 cm^{-1}, and the epoxy groups by the peaks at the area of approximately 887–932 cm^{-1} [30] (Fig. 1.8).

Cadmium (Cd) is a heavy metal and highly poisonous environmental and human health enemy, which is problematic to detect in environmental and analytical research. However, considering the negative consequences, the intervention is highly needed. In recent years, electrochemical techniques have drawn remarkable consideration in the determination of Cd(II) ions because of their privileged benefits. The experiment data indicated that the device is reliable towards the detection of Cd(II) with a limit of detection (LOD) of 45.54 μM and a limit of quantification (LOQ) of 138.00 μM. The sensor had a linear range of output current of 100 -700 lM Cd(II) with a correlation value (R 2 = 0.996), [31].

Here we have modelled some single metal atom catalysts (Al-SWCNTs, Zn-SWCNTs, Fe-SWCNTs) on single wall carbon nanotubes by hand and performed density functional theory calculations to estimate the catalysis of producing hydrogen peroxide and we add many molecules of water and oxygen as reactants. Three various chemical reactions are treated in detail the optimization of the activated reaction molecules, transition states and products are performed. They calculate their activation energies on the basis of which the

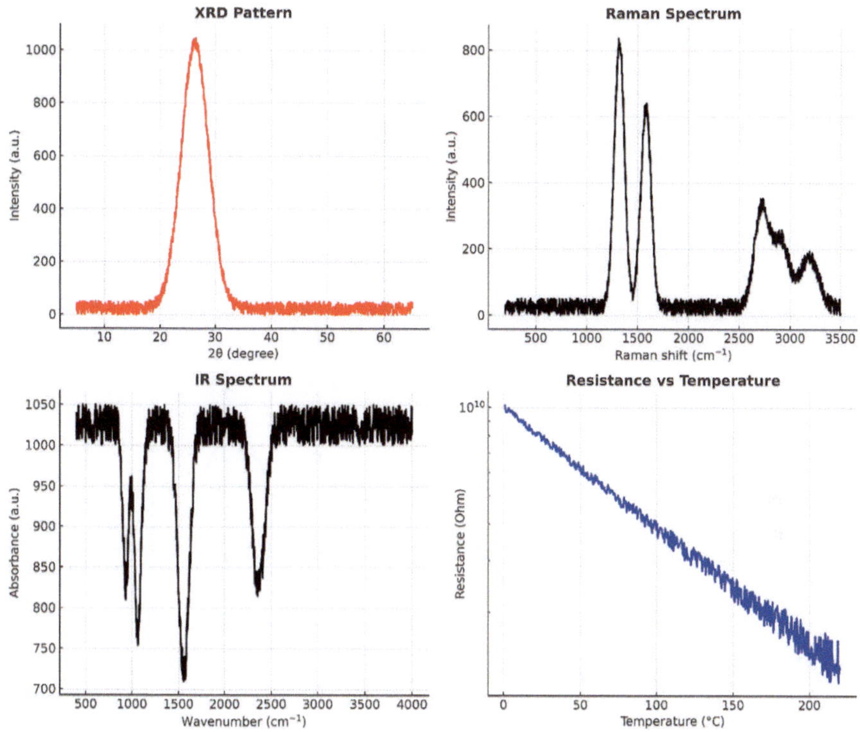

Fig. 1.8 Some properties of graphene oxide

desired reaction pathway can be obtained and possible side products can be checked. Such physical properties can be correlated with their catalytic powers. In addition to the fact that our results compare to the previously performed experiments, which also employed the use of Zn-SWCNTs as catalysts to promote the process, our results also provide us with an idea of the catalytic behavior of Al-SWCNTs and Fe-SWCNTs relative to Zn-SWCNTs [32] (Fig. 1.9).

One of the areas of research is the distribution of SWCNTs in any matrices. The high aspect ratio and tendency of SWCNTs to form agglomerates SWCNTs is unfavorable since a homogeneous dispersion of SWCNTs in composite materials is prefers. Recently, it was recommended that biopolymers such as carboxymethyl cellulose (CMC) could enhance the dispersibility of SWCNTs significantly and additional improvements in the mechanical property of composites [33]. The secret to the introduction of materials in high-value applications such as flexible electronics, and light-weight structural in aerospace engineering lies in the custom process. The SWCNTs beauty is witnessed through its application in energy storage gadgets, including super capacitors and batteries.

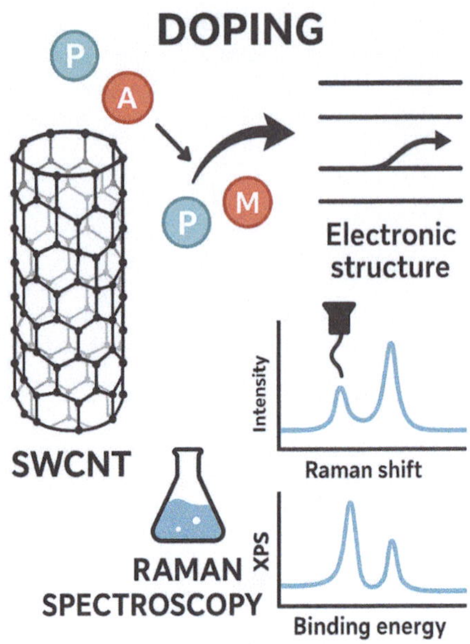

Fig. 1.9 Schematic Representation of doping in single-walled carbon nanotubes (SWCNTs) and corresponding spectroscopic characterization

One of the most considerable findings is associated with the nitrogen doping of SWCNTs that have demonstrated the improvement of the charge storage capacity through the increase of capacitance and electrochemical activity [34].

These properties are basic to the realization of the next-generation supercapacitors which demand materials with larger energy densities and which can be charged and discharged at high rates. Thermoelectric prospects of the SWCNTs are also being achieved with extremely high Seebeck coefficients and enormous electrical conductivities. The thermoelectric of SWCNTs could be tuned by changing the ratio between metallic and semiconducting tubes in a composite, which holds a powerful influence on the overall thermoelectric figure of merit [35]. Single-walled carbon nanotubes (SWCNTs) represent the highly efficient nanodrugs in cancer treatment but the problem is the efficiency of the cellular uptake of these molecules, the cellular fate orientation and the predominant role of the type of the capping molecule that do not permit the application of SWCNTs in the clinics. Of interest is their electrical behavior with regard to their chirality.

Using the opportunity provided by the peculiar biochemical measurements provided by PCA and utilized by K-means, we demonstrate that the concentration of SWCNTDNA happens in two time periods. The initial processes would be (i) transport of the SWCNT-DNA by the vesicles and (ii) re-distribution in the cytoplasm which is followed by presence of two populations of SWCNT-related. The second process starts upon 18 h of cell–cell and SWCNT-DNA interaction. PCA reveals the presence of two SWCNTs-related PC loadings, where new cluster of SWCNTs emerges in the spectra with sharpened and shifted G-peak. It could be attributed to the fact that DNA capping and SWCNTs aggregation do not occur in an acidic environment of autolysosomes due to the fusion of transport vesicles with lysosomes. SWCNTs enter every portion of cells and cause death after 42 to 66 h. SWCNT-ON on its part does not pile up significantly into the glioma cells and at the end of the 72 h exposure period, the SWCNTs accumulation only goes through stage one and never gets in to the second stage.

Great care must be taken with biocompatibility and ecological effect of SWCNTs. Furthermore, the comparative analysis of Sc-SWCNTs with large diameters and with small diameters showed that large-diameter Sc-SWCNTs (enhanced Direct Injection Pyrolytic Synthesis) are superior enhancers as compared to small-diameter Sc-SWCNTs (e.g., HiPCO-based). It is anticipated that this piece of work will introduce the notion of the effective use of Sc-SWCNTs in OSCs and the preparation of the appropriate Sc-SWCNTs semiconductor devices. Most recent findings show that although SWCNTs exhibit toxicity in specific conditions, a modification of their surface chemistry can help reduce adverse outcomes, and therefore enhance biosafety in biomedical practices [36].

We hope that our findings will result in the tremendous amelioration of such nanomaterials usage in medical practice. Besides this, the bio-functionality of SWCNTs is also becoming interesting, especially in biomedicine where they are being investigated with a perspective of administration of drugs and imaging. Their size and the opportunity to chemically modify the surface allow the introduction of bioactive substances, thus, increasing the targeting and therapeutic effect of drugs [37]. Not only was it discovered that functionalized SWCNTs can interact with biological systems but also demonstrated that they can be used in diagnostics and therapies (Fig. 1.10).

The initial maximum is an intensive peak, and the angular location is 2theta = 26.30, and the intensity is I = 8400. A study into this pattern reveals that material originated is a crystalline pattern that is hexagonally Sympathetic. The second, lower Intensity peak angular position is 2theta = 44.50 and the intensity I = 450. The carbon nanotubes synthesized through arc discharge process are morphology-sensitive attributes of the sample exemplified in the diffraction peaks of the XRD. The height of these peaks gives the initial hint of the carbon nanotube diameter as well as the distance between the walls. The X-ray phase analysis of the carbon nanotube gives two maxima one strong maximum at 2 = 26.30 with the intensity I = 1550 and the other weak maximum at 2 = 42.50 with intensity I = 480. The diffraction curve which represents X-ray phase analysis of the carboxyl functionalized carbon nanotube shows that the diffraction spectrum has graphite like features. The spectra are made of two maxima. The former is more intense and they are angular positioned at 2theta = 26.50 with an intensity of I = 1950. The next peak is not so dramatic, and its angular position is 2 = 43.30 and I = 550 [3] (Fig. 1.11).

To be able to use them in healthcare technology, the interactions of SWCNTs with biological environments should be known. The developments that have been made towards the production of SWCNTs continue to trigger the cogitations in a number of directions. The prospects of SWCNTs are limitless in the future that starts with energy efficient devices up to better structural material. Their integration into the smart technologies, e.g.,

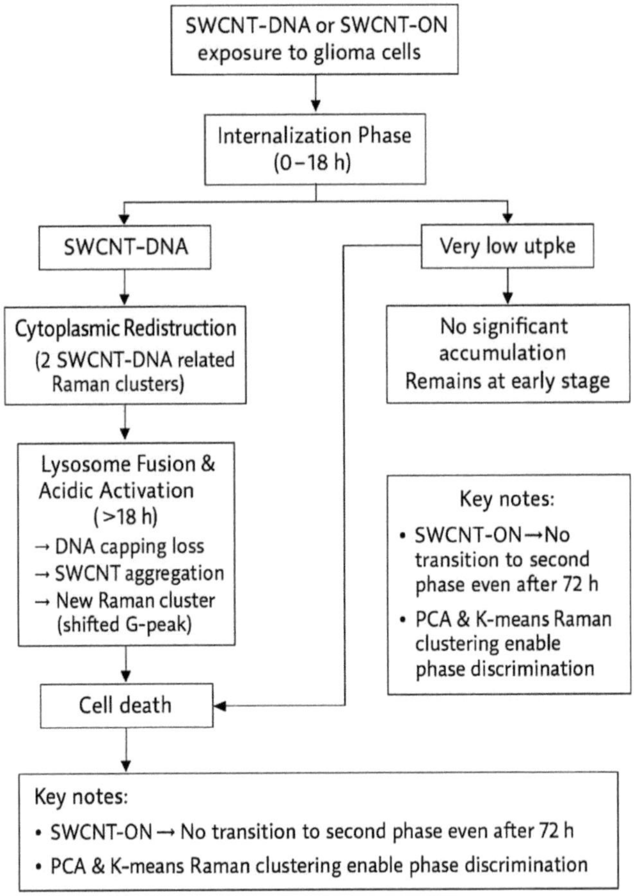

Fig. 1.10 Mechanism of SWCNT accumulation and processing in glioma cells

the sensors and actuators, are already silencing about their abilities to provide the huge contribution to the development of nanotechnology and materials science. It needs the combined efforts of chemistry, physics and materials science communities to realize the full promise of SWCNTs. The issues related to synthesis, functionalization, and applications are being worked on by the researchers in order to produce the new generation of materials which makes use of the magnificent properties of SWCNTs. Studies on their natural behaviors and interactions would yield more applications that will be useful and perhaps even modify how we use nanotechnology to solve most of the problems encountered today. To sum up, SWCNTs compose a considerable field of research and technology that has its unique characteristics specified by the peculiarities of the structure and methods of its synthesis. These achievements in doping, dispersion and composite production emphasize a multi-functional nature of the material in a number of fields. Finally, recent

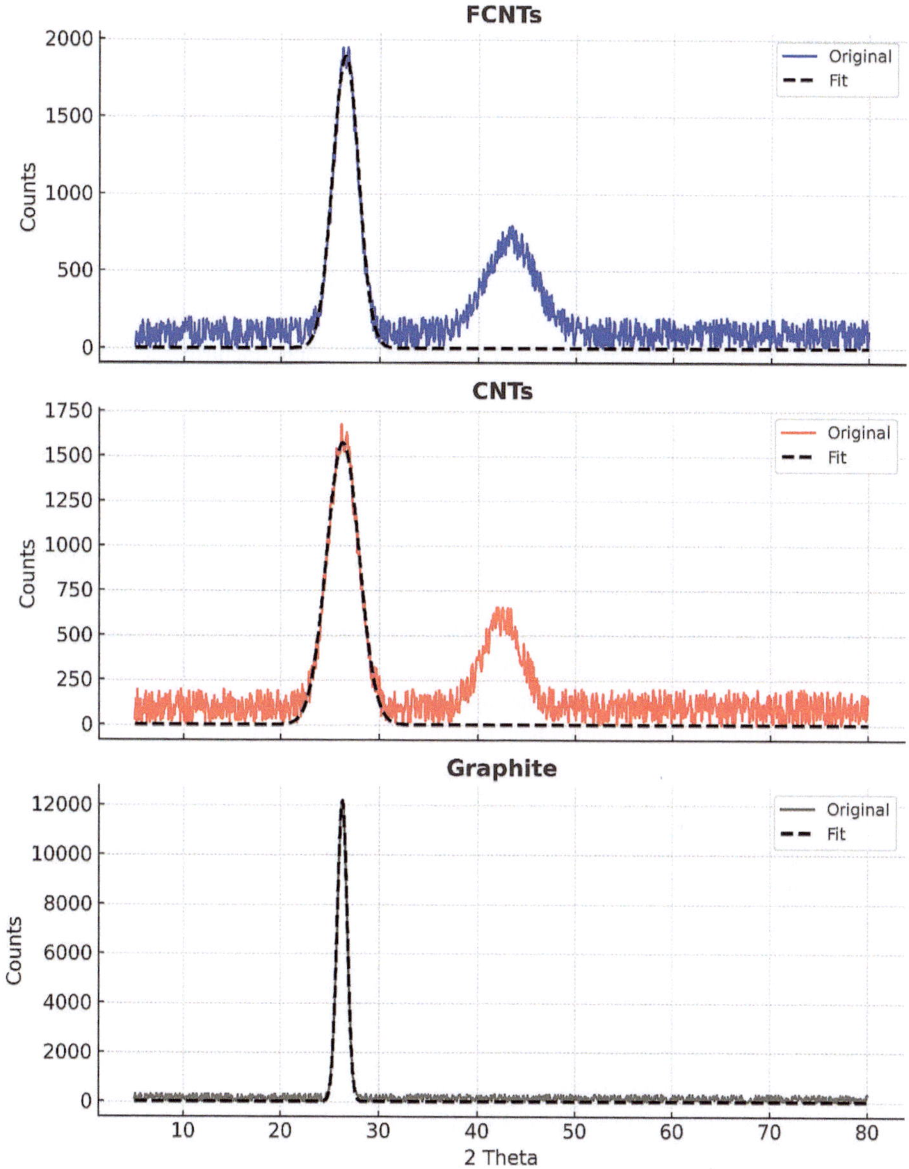

Fig. 1.11 XRD patterns of functionalized CNTs (FCNTs), CNTs, and Graphite

studies into the environmental impact, biocompatibility and functionalization of SWCNTs will carry far reaching consequences in terms of future applications.

1.3 Introduction to MWCNTs

MWCNTs consist of several concentric cylinders of graphene sheets whose inter-layer spacing is enjoying graphite-like configuration (~0.34 nm). The common outer diameters are found between 2 to 100nm, but the length may be many tens or hundreds of micrometers long. Multi-walled carbon nanotubes have excellent mechanical properties, chemical stability and thermal conductivity. Their somewhat low manufacturing cost and ease of processing relative to SWCNTs have made them useful in many structural reinforcements, conductive coatings, supercapacitors, batteries and field emission devices (Table 1.4).

In a systematic manner we experimentally ascertain the effects of the parameters of UV laser-based manufacturing on the morphology and structure of hierarchical microstructures. Optimized hierarchical microstructures soft pressure sensor has a large sensitivity (exceeds 15 kPa^{-1}), and good linearity (R_2 = 0.994) over a range of 0 to 200 kPa. It shows a reaction and recovery time of 57 ms and 62 ms respectively and a commendable stability, as it can withstand more than 5,000 cycles. The shape memory polymers actuation by the laser induces contraction which offers a convenient path in the development of the hierarchical microstructures that can substantially enhance the utility of the flexible pressure sensors in the smart robots and the wearable health care devices. MWCNTs have a high aspect ratio, which, in combination with high electrical conductivity, allows tailoring these materials to extremely different fields, including electronics, medicine, or environmental research [38].

Table 1.4 Principal structural and physical properties of multi-walled carbon nanotubes

Property	Description
Structure	Multiple concentric graphene cylinders
Diameter	~2–100 nm
length	Tens to hundreds of micrometers
Interlayer spacing	~0.34 nm (similar to graphite)
Electrical conductivity	High; less sensitive to chirality
Mechanical strength	Superior to SWCNTs in bulk composites
Thermal conductivity	High (~3000 W/m·K)
Chemical Stability	Excellent chemical and thermal stability
Typical applications	Structural reinforcement, energy storage devices, field emission devices

The findings revealed that synergistic interactions played a key role in achieving the distinctive X-ray attenuation characteristics of the material. The linear attenuation coefficient of the fabricated bandage—comprising untreated cotton wrapped with a CeO_2/MWCNT/PDMS composite—ranged from 1.274 m^{-1} to 0.549 m^{-1}, while the mass attenuation coefficients were found to vary between 0.823 $m^2 \cdot kg^{-1}$ and 0.354 $m^2 \cdot kg^{-1}$ within the energy range of 40 to 100 kVp. The binary protective effect of CeO_2 nanoparticles and MWCNT leads to the improvement in the key properties of high density and protection efficiency [39].

Multi-Walled Carbon Nanotubes include several concentric graphene cylinders. Interlayer spacing:

$$d_{interlayer} \approx 0,34nm$$

The diverse phenomena about MWCNTs can be attributed to the synergistic effects among its concentric layers:

- The electrical conductivity is roughly the sum of the contribution of each of the layers.
- Van der Waals force between layers gives improvement in mechanical strength.

Young (E) of Multi-Walled Carbon Nanotubes (MWCNTs): $E \approx 0,9 - 1,2TPa$.

Thermal conductivity (axial):

$$k \approx 3000 - 3500 \frac{W}{mK}$$

Electrical conductivity:

$$\sigma \approx 10^5 - 10^6 \frac{S}{m}$$

MWCNTs are normally synthesized by chemical vapour deposition, arc discharge and laser ablation methods and this enables synthesis of MWCNTs of varying diameter and length and hence manipulation of their properties to suit particular applications. Multiwalled carbon nanotubes (MWCNTs) were subjected to oxidation and acidification using concentrated nitric and sulfuric acids (HNO_3 and H_2SO_4), resulting in modified MWCNTs (M-MWCNTs). Both the raw and functionalized nanotubes were characterized through a range of analytical techniques. The study focused extensively on their performance in demanganization applications. Experimental findings demonstrated that the M-MWCNTs exhibited a significantly enhanced adsorption capacity for Mn (II), increasing from approximately 20% to 75% following surface modification. The outcomes of the research suggest a potential method of optimization of the manganese removal in wastewater; simultaneously, the use of different mathematical models allows to gain more scientific proofs on the processes of the heavy metals' adsorption by nanomaterials [40].

Their compatibility with other substances may be enhanced by treating the MWCNT surfaces, e.g. with carboxyl groups or amino groups, and thus the opportunity to incorporate them into composites, which require improved interfacial contacts and bonding. The sensor's electrode surface was modified with amine-functionalized reduced graphene oxide–Fe_3O_4 nanoparticles (MRGO-NH_2), which served as the electroactive sensing platform. Electrochemical signal of the MRGO-NH_2 platform material also appeared compared to the reduced graphene oxide. In the name Pt-Fe3O4-MWCNTs-NH_2, the multiwall carbon nanotubes were used as a support to deposit a high catalytic loading of Pt and Fe_3O_4 nanoparticle. The fabricated immunosensor was also characterized by high specificity, acceptable repeatability and stability besides. The sensor could be used to analyze sample of breast cancer patients and gave reproducible results [41].

The currently acknowledged and most frequently used medicines, aimed at the treatment of severe depression and related disorders are the selective serotonin reuptake inhibitors. This paper gives a brief overview of recent progress in making electrochemical sensors to detect selective serotonin reuptake inhibitors. According to published studies, composites and nanocomposites on modified electrodes are more common in the development of sensors in the selective serotonin reuptake inhibitors detection than single materials due to the synergistic effects that enhance detection. Studies have demonstrated carbon-metal and carbon-polymer composite are the most common functional materials used to modify electrode surfaces. carbon-metal nanocomposites multi-walled carbon nanotubes (MWCNTs) were predominantly employed in carbon-metal nanocomposites along with metal and/or metal oxide nanoparticles. Such materials had shown extraordinary detection limits and high selectivity owing to synergistic effects that enhance better catalytic reaction on the target analyte. The best sensitivity to the measurement and detection of SSRIs was observed with metal-polymer composites, and the limit of detection was as low as picomolar. Metal electrodes are rarely applied, as they enable an increase of sensitivity and selectivity and reduce the interference and therefore, are suited in clinical practice (Fig. 1.12).

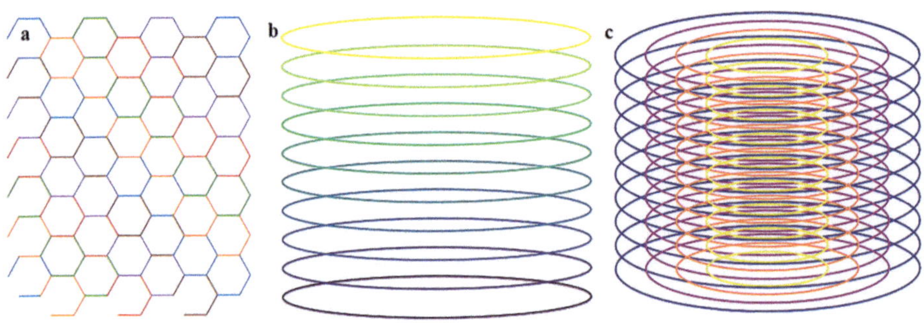

Fig. 1.12 Schematics diagrams in graphene list (**a**), SWCNTs (**b**), MWCNTs (**c**)

The further development of new materials and nanomaterials is required to improve surface functionality of the electrode, to enable ultra-trace detection of selective serotonin reuptake inhibitors in various types of samples and thereby increase convenience. Electrode surface modification with functional materials has shown great advantages and potential of selective serotonin reuptake inhibitors sensing. Besides, MWCNTs play a crucial role in electrochemical purposes, especially as electrode materials owing to their high surface area as well as conductivity [42] (Fig. 1.13). The electrochemical characteristics of the composite material enforced with MWCNTs allow sensitive detection of biomarkers of physiological interest in a clinical environment [43].

It has been found that MWCNTs reinforcement in polymer matrices could considerably improve mechanical and thermal properties, and therefore, this expands their applicability in lightweight and high-strength materials in aerospace and automotive industries [44, 45].

Besides this, MWCNTs can be employed in the remediation of the environment since they can absorb heavy metals in a contaminated water body owing to their high surface area and contaminant preference. This capability enhances the purity of water and the reason why MWCNTs are versatile in their application on environmental issues. Their functionalization assists them in maximizing their adsorption performance and selectivity

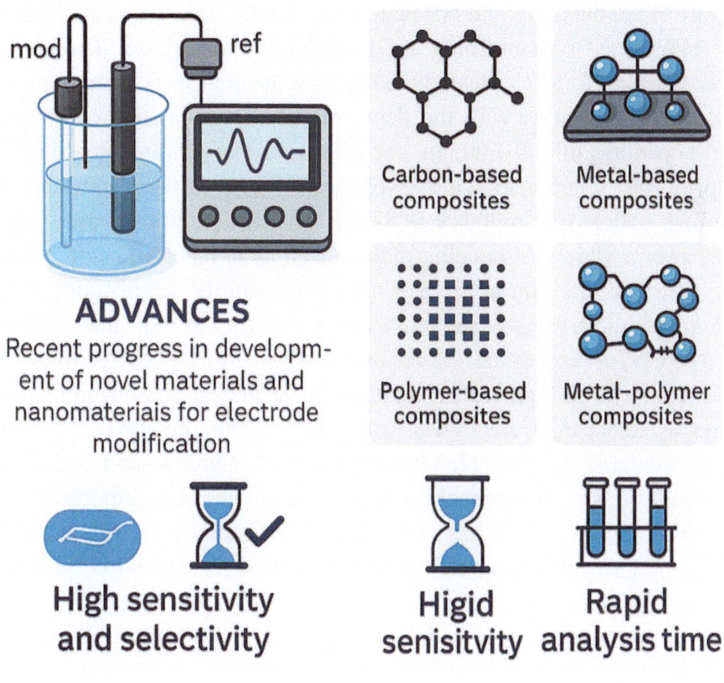

Fig. 1.13 Electro-chemical sensors for selective serotonin reup-take inhibitors detection

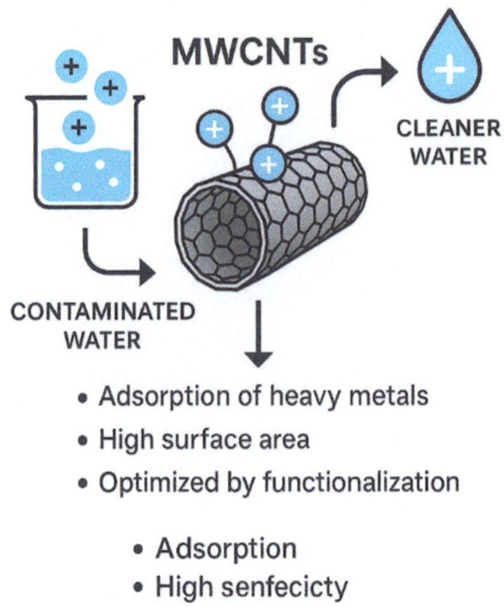

Fig. 1.14 Applications of multi-walled carbon nanotubes (MWCNTs)

to broad collections of pollutants in achieving sustainable environmental practices [46] (Fig. 1.14).

The electrical conductivity and adequate heat transfer capacity of multiwalled carbon nanotubes (MWCNTs) is astonishing and therefore, to achieve an ideal combination of these properties, they form an ideal candidate as a material in flexible transparent films. The flexible MWCNT films with the depth of the cavity cavities of between 23 and 100 nm, grid openings of 100–600 nm, and grid widths of 6–80 nm could reach the temperature up to 106 °C after applying a voltage of 30 V. At the same time, these films provided a high optical transmittance as 87.6%. The flexible films possess high mechanical stability over a variety of bending radii as well as in repeated cycle bending fatigue tests. More than that, the film allows not only to control the temperature accurately, but also to see the characteristics of the temperature variation in real time in illumine scent materials. MWCNTs exceptional are advanced when utilized in electronics as flexible components. Due to their high electrical conductivity, they are finding application in flexible displays as transparent electrodes, solar cells, highlining their significance in the realization of wearable and portable electronic devices. Clothes form one of the necessities of human existence. The annual production of clothes is approximately 100 billion items. The proliferation of the IoT technology is expected to result in the worldwide spending on smart wearable gadgets totaling up to US$ 23.82 billion by the end of the year 2031.

Convergence in industries (industrial convergence) between electronics and textiles is occurring at a high rate in order to bring on-body platform to seamless computing and does not compromise wonderful textile characteristics such as excellent comfort, flexibility, wash-ability, mechanical strength, air permeability, light-weight and fold-ability.

T-TENG may be vital in the field of smart wear. T-TENG is also capable of serving as a power source and wearable devices become energy independent as it does not require batteries. It was discovered that carbonaceous nanofillers played a very useful role in enhancing the performance of these T-TENGs as either the triboelectric or electrodes material. Within the scope of the carbonaceous nanofillers, 1D and 2D, and its derivatives have the most significant role when it comes to triboelectric nanogenerators. The report offers design principles over the emerging materials, the underlying operational regimes, flexible textile structure on TENG, various fabrication strategies that can be utilized in integration of carbonaceous materials and the efficacy of the commonly utilized low dimensional carbonaceous nanofillers in T-TENGs. T-TENGs would find excellent applications in self-powered systems, wearable sensors, identification verification and health monitoring. Weaving is the most frequent textile structure of the symmetrical contact surface with great dimensional stability, durability, strength, and surface area but thinner thickness, fitting CS, FS, and LS mode-based T-TENGs. The benefit of knitting is that the contact surfaces are asymmetrical because of the benefits of elasticity and easier deformation, which is promising in SE and CS mode-based T-TENGs.

This implies that knitted fabrics pill, snag and shrink highly and are hence not used in FS or LS modes. Nonwoven and braiding T-TENGs are hardly developed because of their structural faults. fabric-made TENGs have greater output efficiency, but they exhibit inadequate wear resistance. but fiber-based TENGs, on the other hand, can be made more deformable but not compatible with current manufacturing processes. Contact electrification may be potential by nanopatterning using carbonaceous nanofillers and enhanced by three-dimensional fabric structures, and irreproducible. Simultaneously, energy, porosity, geometric shape, and elevate polarity, all of which are beneficial in enhancing the efficiency of energy conversion of T-TENGs. The revealed T-TENGs were compared according to the performance, and the problems that may arise in the future were indicated. The carbons material, textiles, and TENGs synergies were successfully revealed in this article. With the help of this review, researchers working in various spheres, including materials, textiles, and electronics, will profit. The results can inform the smart clothing producers in the appropriate choice of triboelectric, electrode, and carbonaceous materials and manufacturing techniques when preparing and commercializing their T-TENGs. The new studies observe their addition into new materials that can generate energy via triboelectric nanogenerators and demonstrate potential in smart textiles and other wearable technologies [47, 48] (Fig. 1.15).

It appears in the form of a graphic account of the main working regimes, textile topologies, the significance of carbon nanofillers, interactions, and opportunities of textile-based triboelectric nanogenerators (T-TENGs). Other alternatives of carbon nanofillers including MWCNTs, graphene and carbon black enhance surface properties and increase conversion of energy efficiency. But biocompatibility and toxicity of MWCNTs, especially long-term impacts on human health and environment are raising concern. Their toxicological profile investigations reveal a possible danger with regard to both inhalation

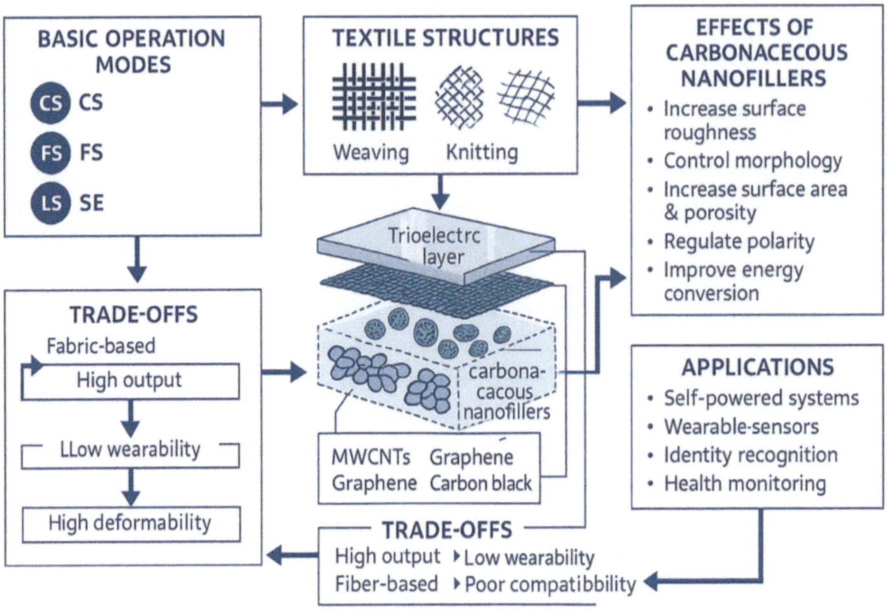

Fig. 1.15 Schematic overview of textile-based triboelectric nanogenerators (T-TENGs) enhanced with carbonaceous nanofillers

and systemic exposure, making it necessary to conduct a rigorous safety testing prior to broad implementation [49, 50]. The ongoing research on the safety, functionalization and usage of the MWCNTs is highly significant to streamline their economic viability without compromising the well-being of the population and the environment (Table 1.5).

They have applications in advanced material in nanoelectronics and environmental remediation and challenges such as biocompatibility and safety need to be carefully overcome to achieve their potential. With more understanding about MWCNTs, more innovations concerning their synthesis, functionalization, and application will definitely push the boundaries of the existing technology.

1.4 Concluding Remarks

The classification of CNTs is significant in the utilization of the unique property in many fields of technology. The diversity of SWCNTs and MWCNTs and their sub-classes suggests the flexibility of CNTs as graceful functional materials. As the synthesis methods are being perfected and allow to exert more control over the chirality, purity, and integrity of the structure, it is rather likely that the range of applicability of CNT will be broadened,

Table 1.5 Comparative summary of SWCNTs and MWCNTs

Feature	SWCNTs	MWCNTs
Structure	Single graphene cylinder	Multiple concentric cylinders
Diameter	~0.4–2 nm	~2–100 nm
Length	Several micrometers	Tens to hundreds of micrometers
Electronic Behavior	Metallic or semiconducting (chirality-dependent)	Predominantly metallic
Mechanical properties	High tensile strength and flexibility	Very high bulk strength
Thermal conductivity	Very high	High
Typical cost	Higher (difficult synthesis)	Lower (easier mass production)
Main applications	Electronics, sensors, biomedical	Composites, energy storage, coatings

triggering new discoveries, both, in the industrial and academic domains. Their categorized knowledge in classification and the related physical equations help to have a finer control in their preparing and adapting to the specific use.

Classification Carbon nanotubes (CNTs) can be classified in an inclusive scheme encompassing structural, dimensional, electrical, and functional properties. Their degree of functionalization, walls, chirality, aspect ratio, synthesis route have a significant impact on physicochemical properties of CNTs and, thereby, on their applicability in the broad spectrum of fields. The tunable electrical properties of SWCNTs, and MWCNTs their outstanding mechanical flexibility have remained of importance in nanoelectronics, sensing and medicinal applications. Doping techniques, dispersion techniques, and composite embedding show advancement and flexibility, which implies their ability to innovate. MWCNTs have been touted to have an exceptional mechanical strength, chemical stability along with the possibility of large-scale production rendering them essential to future applications of advanced structural materials, conductive composites and environmental technologies (Table 1.6).

However, serious concerns with regard to biocompatibility, toxicity and environmental effects after long term use must be clarified acceptably to enable their use to be liberalized. As the research advances further, a constant growth in the preciseness of the synthesis, the technique of functionalization and composite architecture would be anticipated to contribute to the further enhancement of the applications of CNT in the scientific as well as the industrial community. In this dynamic climatic condition, classification necessity and relevance of structure-property interactions of CNTs will need to be known so as to gain maximum out of such materials and define the future direction of nanotechnology.

Table 1.6 Summary table of key equations

Property	Formula/Expression
Chiral vector	$\overrightarrow{C_h} = n\overrightarrow{a_1} + m\overrightarrow{a_2}$
Tube diameter d_t	$d_t = \frac{a}{\pi}\sqrt{n^2 + m^2 + nm}$
Chiral angle θ (theta)	$\theta = \tan^{-1}\left(\frac{\sqrt{3}m}{2n+m}\right)$
Band gap of SWCNTs	$E_g = \frac{2\gamma_0\alpha}{d_t}$
Density of states (DOS)	$D(E) \propto \frac{1}{\sqrt{E-E_n}}$
Interlayer spacing in MWCNTs	$d_{interlayer} \approx 0,34nm$
Young's modulus of MWCNTs	$E \approx 0,9 \div 1,2TPa$
Thermal conductivity (MWCNTs)	$k \approx 3000 \div 3500\frac{W}{mK}$
Electrical conductivity (MWCNTs)	$\sigma \approx 10^5 \div 10^6\frac{S}{m}$

References

1. Abaszade, R. G., Kapush, O. A., & Nabiyev, A. M. (2020). Properties of carbon nanotubes doped with gadolinium. *Journal of Optoelectronic and Biomedical Materials*, 12(3), 61–65. https://www.chalcogen.ro/61_AbaszadeRG.pdf
2. Abaszade, R. G., Kapush, O. A., Mamedova, S. A., Nabiyev, A. M., Melikova, S. Z., & Budzulyak, S. I. (2020). Gadolinium doping influence on the properties of carbon nanotubes. *Physics and Chemistry of Solid State*, 21(3), 404–408. https://doi.org/10.15330/pcss.21.3.404-408
3. Stetsenko, M. and Abaszade, R. (2023). X-ray phase analysis of carbon nanotubes obtained by the arc discharge method. UNEC j. eng. appl. sci., 3(1), 15–20. https://doi.org/10.61640/ujeas.2023.0503
4. Niu, X., Liu, Y., Zhao, R., Wang, L., Yuan, M., Zhao, H., Li, H., & Wang, K. (2024). Chiral carbon nanostructures: a gateway to promising chiral materials. Journal of Materials Chemistry A, 12(28), 17073–17127. https://doi.org/10.1039/d4ta02396j
5. Yaqoob, U., Lenz, W., Alcheikh, N., Jaber, N., & Younis, M. (2022). Highly selective multiple gases detection using a thermal-conductivity-based mems resonator and machine learning. Ieee Sensors Journal, 22(20), 19858–19866. https://doi.org/10.1109/jsen.2022.3203816
6. Abbasi, R., Shineh, G., Mobaraki, M., Doughty, S., & Tayebi, L. (2023). Structural parameters of nanoparticles affecting their toxicity for biomedical applications: a review. Journal of Nanoparticle Research, 25(3). https://doi.org/10.1007/s11051-023-05690
7. Lima, E., Martins, G., Diaz, R., Schechter, M., Piqueira, J., & Justo, J. (2022). Effects of carbon nanomaterials and aloe vera on melanomas—where are we? recent updates. Pharmaceutics, 14(10), 2004. https://doi.org/10.3390/pharmaceutics14102004
8. Senthilkumar, V. (2022). Some approximate buckling solutions of triple-walled carbon nanotube. Vietnam Journal of Mechanics. https://doi.org/10.15625/0866-7136/17054
9. Oliveira, C. and Rocha, V. (2023). Effective quantum graph models of some nonequilateral graphyne materials. C – Journal of Carbon Research, 9(3), 76. https://doi.org/10.3390/c9030076

10. Muchlisha, N., Widjonarko, D., & Saraswati, T. (2023). Synthesis of drug carrier carbon nanofoam by chemical vapor deposition using agar/nacl catalyst. Journal of Physics Conference Series, 2556(1), 012006. https://doi.org/10.1088/1742-6596/2556/1/012006
11. Ding, Y., Liu, B., Li, M., Hao, L., Zhu, Z., Han, Y., Ma, F.Y., & Xiang, Z. (2022). High performance copper-tungsten @graphene composites materials. Journal of Physics Conference Series, 2321(1), 012010. https://doi.org/10.1088/1742-6596/2321/1/012010
12. Wei, J. (2023). Research status of lithium oxygen battery and its components. Highlights in Science Engineering and Technology, 43, 499–503. https://doi.org/10.54097/hset.v43i.7469
13. Taha, Z. and Major, A. (2023). A review on mwcnts: the effect of its addition on the polymer matrix. Gradus, 10(1). https://doi.org/10.47833/2023.1.eng.012
14. Raul, P., Thakuria, A., Das, B., Devi, R., Tiwari, G., Yellappa, C., & Kamboj, D. (2022). Carbon nanostructures as antibacterials and active food-packaging materials: a review. ACS Omega, 7(14), 11555–11559. https://doi.org/10.1021/acsomega.2c00848
15. Musa, N. (2023). Carbon nanomaterials and their applications., 40–71. https://doi.org/10.21741/9781644902172-3
16. Amadi, E., Venkataraman, A., & Papadopoulos, C. (2022). Nanoscale self-assembly: concepts, applications and challenges. Nanotechnology, 33(13), 132001. https://doi.org/10.1088/1361-6528/ac3f54
17. Yatoo, M., Habib, F., Husain, A., Ahmad, S., & Ahmad, Z. (2023). A concise review of single-molecule magnets, carbon nanotubes and hybrids between them. https://doi.org/10.20944/preprints202306.1276.v1
18. Asri, N., Saraswati, R., Yogaswara, R., Restyani, N., & Suprapto, S. (2022). Converting of kesambi (schleichera oleosa l.) oil into biodiesel using zno-based solid acid catalyst. Iop Conference Series Earth and Environmental Science, 1034(1), 012054. https://doi.org/10.1088/1755-1315/1034/1/012054
19. Wang, B. (2024). Preparation of carbon nanotube transistors and their applications in integrated circuits. Highlights in Science Engineering and Technology, 111, 144–147. https://doi.org/10.54097/b45htv75
20. Xu, Q., Ouyang, H., Lin, Y., Ye, M., Wang, X., Sun, L., Chen, T., & Ma, L. (2023). Welding model for a carbon nanotube bundle with staggered electrodes: implications for conductivity with joule heating. ACS Applied Nano Materials, 6(15), 14488–14497. https://doi.org/10.1021/acsanm.3c02597
21. Badran, A., Uthumporn, U., Yussof, N., & Bashir, M. (2023). Advancements in adsorption techniques for sustainable water purification: a focus on lead removal. Separations, 10(11), 565. https://doi.org/10.3390/separations10110565
22. Nashed, O., Partoon, B., Lal, B., Sabil, K., Al-Arabi, J., & Bavoh, C. (2024). Kinetic study and CO_2 separation assessment from a CH_4/CO_2 gas mixture in the presence of multiwall carbon nanotubes. Energy & Fuels, 38(23), 22986–22996. https://doi.org/10.1021/acs.energyfuels.4c04125
23. Goci, M., Taka, A., Martin, L., & Klink, M. (2023). Chitosan-based polymer nanocomposites for environmental remediation of mercury pollution. Polymers, 15(3), 482. https://doi.org/10.3390/polym15030482
24. Adhikari, C., Morris, D., Noonan, T., Neupane, T., Lamichhane, B., & Gautam, B. (2023). Dispersion in single-wall carbon nanotube film: an application of bogoliubov–valatin transformation for hamiltonian diagonalization. https://doi.org/10.22541/au.167809331.13364369/v1
25. Luan, J., Duan, W., Li, Y., Meng, F., Liu, Y., Zhang, X., Zhou, J., Li, W.-Z., & Fu, Y. (2023). Metal–organic framework as catalyst precursor of floating catalyst chemical vapor deposition for single-walled carbon nanotube manufacture. ACS Sustainable Chemistry & Engineering, 11(33), 12423–12434. https://doi.org/10.1021/acssuschemeng.3c03147

26. Everhart, B., Rao, R., Nikolaev, P., Liu, T., Gómez-Gualdrón, D., Maruyama, B., & Amama, P. (2022). High-throughput experimentation for selective growth of small-diameter single-wall carbon nanotubes using ru-promoted co catalysts. Chemistry of Materials, 34(10), 4548–4559. https://doi.org/10.1021/acs.chemmater.2c00347
27. Abaszade, R. G., Mammadova, S. A., Kapush, O. A., Agayev, F. G., Nabiev, A. M., Mammadova, M. M., Budzulyak, S. I., & Kotsyubynsky, V. O. (2021). Synthesis and characterization of graphene oxide flakes for transparent thin films. *Physics and Chemistry of Solid State*, 23(3), 595–601. https://doi.org/10.15330/pcss.22.3.595-601
28. Abaszade, R. G., Aliyev, E. M., Babanli, M. B., Kotsyubynsky, V. O., Zapukhlyak, R. I., Mammadov, A. G., Budak, H. F., Kasapoglu, A. E., Gür, E., Margitych, T. O., & Stetsenko, M. O. (2023). Investigation of thermal properties of carbon nanotubes and carboxyl group-functionalized carbon nanotubes. *Physics and Chemistry of Solid State*, 24(3), 530–535. https://doi.org/10.15330/pcss.24.3.530-535
29. Guliyeva, N. A., Abaszade, R. G., Khanmammadova, E. A., & Azizov, E. M. (2023). Synthesis and analysis of nanostructured graphene oxide. *Journal of Optoelectronic and Biomedical Materials*, 15(1), 23–30. https://doi.org/10.15251/JOBM.2023.151.23
30. Abaszade, R. G. (2022). Synthesis and analysis of flakes graphene oxide. *Journal of Optoelectronic and Biomedical Materials*, 14(3), 107–114. https://doi.org/10.15251/JOBM.2022.143.107
31. Sharma, B., Singh, A., Sharma, A., Dubey, A., Gupta, V., Abaszade, R. G., Sundramoorthy, A. K., Sharma, N., & Arya, S. (2024). Synthesis and characterization of zinc selenide/graphene oxide (ZnSe/GO) nanocomposites for electrochemical detection of cadmium ions. *Applied Physics A*, 130(297). https://doi.org/10.1007/s00339-024-07472-0
32. Chen, Y., Zhang, S., Mao, D., Xie, R., Qin, Q., Su, X., Zhai, B., Li, L.-C., & Zheng, Y. (2022). Metals (al, fe, zn) doped in single walled carbon nanotubes for catalytic oxidation of h2o to h2o2: a theoretical investigation. New Journal of Chemistry, 46(33), 15811–15819. https://doi.org/10.1039/d2nj02237k
33. Luo, X., Zhou, L., Wang, J., Chen, Y., & Lu, L. (2024). Crucial role of volumetric variation tolerable single-walled carbon nanotube/polymer binder network for stabilizing bulk silicon anode. ACS Applied Polymer Materials, 6(20), 12441–12450. https://doi.org/10.1021/acsapm.4c01558
34. Dong, H., Zhang, L., Liao, Y., Huang, K., Lian, C., Zhou, X., Zhang, Z., Kauppinen E.I., & Wu, Z. (2023). Floating catalyst chemical vapor deposition patterning nitrogen-doped single-walled carbon nanotubes for shape tailorable and flexible micro-supercapacitors. Advanced Functional Materials, 33(29). https://doi.org/10.1002/adfm.202301103
35. Ma, X., Wang, Y., Wang, C., Zhang, Y., Fu, P., & Du, F. (2024). Enhanced thermoelectric properties of exfoliated bn nanosheets/single-walled carbon nanotube composite films for applications in flexible electronics. ACS Applied Nano Materials, 7(13), 15640–15647. https://doi.org/10.1021/acsanm.4c02648
36. Lin, H., Miyata, D., Yagisawa, M., Huda, M., Hashimoto, S., Hashimoto, T., & Matsuo, Y. (2022). A mechanistic study on enhanced PEDOT:PSS for organic solar cells using semiconducting single-walled carbon nanotubes in different diameters. Applied Physics Express, 15(12), 121001. https://doi.org/10.35848/1882-0786/ac9d80
37. Golubewa, L., Timoshchenko, I., & Kulahava, T. (2024). Specificity of carbon nanotube accumulation and distribution in cancer cells revealed by k-means clustering and principal component analysis of Raman spectra. The Analyst, 149(9), 2680–2696. https://doi.org/10.1039/d3an02078a078a
38. Chen, R., Zhang, C., Xiao, C., Zhao, T., Luo, T., & Zhou, W. (2024). Fabrication of hierarchical microstructures via laser-induced shrinkage of shape memory polymers for flexible pressure

sensing. ACS Applied Materials & Interfaces, 16(34), 45732–45744. https://doi.org/10.1021/acsami.4c09859

39. Verma, S., Dhangar, M., Bajpai, H., Chaturvedi, K., Mohapatra, R., Khan, M., Azam, M., Al-Reseyes, A.I., & Srivastava, A. (2023). X-ray radiation shielding and microscopic studies of flexible and moldable bandage by in-situ synthesized cerium oxide nanoparticles/MWCNTs nanocomposite for healthcare applications. RSC Advances, 13(13), 8594–8605. https://doi.org/10.1039/d3ra00067b
40. Zhou, Y., He, Y., Wang, R., Mao, Y., Bai, J., & Dou, Y. (2023). Modification of multiwalled carbon nanotubes and their mechanism of demanganization. Molecules, 28(4), 1870. https://doi.org/10.3390/molecules28041870
41. Patil, S., Karade, V., Kim, J., Chougale, A., & Patil, P. (2024). Electrochemical detection of a breast cancer biomarker with an amine-functionalized nanocomposite PT–Fe_3O_4–MWCNTs–NH_2 as a signal-amplifying label. ACS Applied Materials & Interfaces, 16(20), 25601–25609. https://doi.org/10.1021/acsami.3c15531
42. Barry, S., Franke, C., Mulaudzi, T., Pokpas, K., & Ajayi, R. (2023). Review on surface-modified electrodes for the enhanced electrochemical detection of selective serotonin reuptake inhibitors (ssris). Micromachines, 14(7), 1334. https://doi.org/10.3390/mi14071334
43. Mirfendereski, S., Ahmari, M., & Gargari, S. (2023). Experimental investigation of the influence of nanofluids on the wetting of membrane contactors in the co2 capture process. Energy & Fuels, 37(24), 19758–19770. https://doi.org/10.1021/acs.energyfuels.3c03968
44. Chen, J., Zhang, T., Xu, Y., Li, H., Cui, H., Zhao, X., Qu, K. & Cui, Z. (2025). Molecularly imprinted electrochemical sensor based on MWCNTs/GQDS for the detection of sulfamethazine in aquaculture seawater. Biosensors, 15(3), 184. https://doi.org/10.3390/bios15030184
45. Lin, X., Wang, W., Chang, X., Chen, C., Guo, Z., Yu, G., Zhou, Y., Qu, K., & Li, H. (2024). ROS/MTROS promotes tnts formation via the Pi3k/aktmtor pathway to protect against mitochondrial damages in glial cells induced by engineered nanomaterials. Particle and Fibre Toxicology, 21(1). https://doi.org/10.1186/s12989-024-00562-0
46. Li, X., Cui, Y., Du, W., Cui, W., Huo, L., & Liu, H. (2024). Adsorption kinetics and mechanism of pb(ii) and cd(ii) adsorption in water through oxidized multiwalled carbon nanotubes. Applied Sciences, 14(5), 1745. https://doi.org/10.3390/app14051745
47. Nie, B., Wang, J., Cui, T., Zhu, Y., Peng, W., Zhu, M., Xi, R., Fan, S., & Wei, R. (2025). High-aspect-ratio multiwalled carbon nanotube structures for flexible transparent electrothermal films. Langmuir, 41(18), 11454–11462. https://doi.org/10.1021/acs.langmuir.5c00434
48. Sayam, A., Rahman, M., Sayem, A., Ahmed, A., & Alimuzzaman, S. (2024). Carbon-based textile-structured triboelectric nanogenerators for smart wearables. Advanced Energy and Sustainability Research, 5(11). https://doi.org/10.1002/aesr.202400127
49. Fukushima, S., Kasai, T., Senoh, H., Umeda, Y., Mine, T., Sasaki, T., Kondo, H., Matsumoto, M., & Aiso, S. (2023). Contribution of toxicological pathology to occupational health: lung carcinogenicity of fibrous and particulate substances in rats. Journal of Toxicologic Pathology, 36(2), 69–83. https://doi.org/10.1293/tox.2022-0086
50. Liu, H., Mao, H., Ouyang, X., Lu, R., & Li, L. (2024). Intercellular mitochondrial transfer: the novel therapeutic mechanism for diseases. Traffic, 25(9). https://doi.org/10.1111/tra.12951

Engineering Properties of Carbon Nanotubes 2

Carbon nanotubes (CNTs) provide an excellent example of nanomaterials because these are materials that have incredible mechanical, electrical and thermal properties with the potential to shift the performance of various composite materials. They are differentiated by the number of layers of a graphene structure to form the tube that attributes to their usefulness used in virtually every nanotechnological and materials science application[1]. Carbon nanotubes distribution and location are significant in order to increase the mentioned properties. A good dispersion in a polymer matrix promotes the mechanical strength, the thermal stability and the electrical conductivity. Distributing the CNTs well will give a percolating network that can improve the transmission of load, along with distribution of stress in the composite [2]. In the present study, white rice husk ash (WRHA) was undergoing various plasma treatment therapies to modify it to be more favorable to act as a reinforcing filler in natural rubber (NR) composites. These treatments were argon (cleaning), Oxygen (activation) and acetylene (polymerization) plasma and the aim of these treatments was to increase the WRHA specific surface area. At that, a set of five rubber formulations on the basis of conventional NR/silica tire tread formulation was synthesized, and the kinetics of vulcanization, mechanical, dynamic and tribological characteristics were determined. In its turn, the sufficiently decreased surface free energy of the modified filler was confirmed by the analysis and preconditioned its more appropriate dispersion and interactions with the NR matrix (Fig. 2.11).

WRHA treated by argon plasma, in contrast to most of the other changes, inhibited vulcanization and reduced the strength of the vulcanizates. This can be attributed to the fact that; all the plasma treatments enhanced the abrasion resistance and lower tan δ values at 60 °C and 100 °C that implies low rolling resistance and minimized heating. WRHA that was subjected to argon plasma treatment and subsequent polymerization using an

R. G. O. Abaszade and A. K. Haghi, *Carbon Nanotube Devices for Nanoelectronics*,
Synthesis Lectures on Solid State Materials and Devices,
https://doi.org/10.1007/978-3-032-02114-4_2

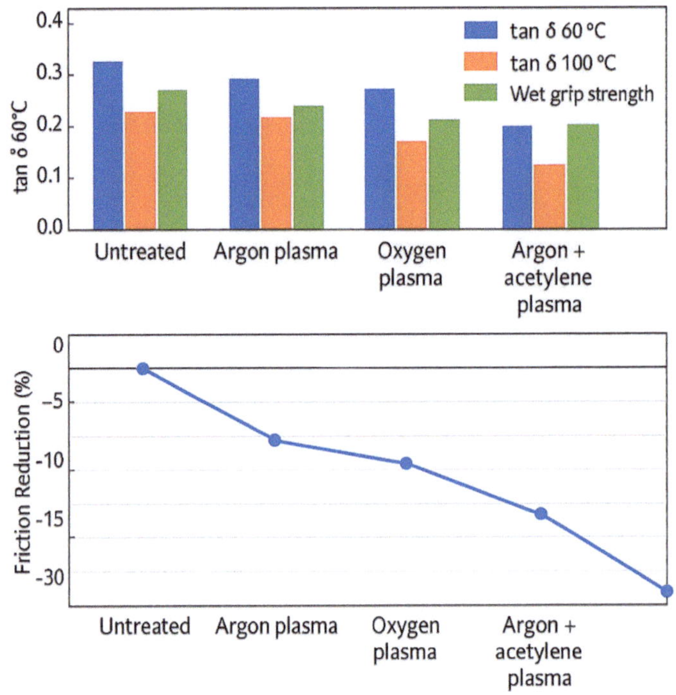

Fig. 2.1 Plasma-treated white rice husk ash as a reinforcing filler in natural rubber

acetylene plasma considerably showed the highest wet grip strength (~44%), lower friction and wear (~14%). These findings demonstrate that plasma treatment especially the two steps treatment is a possible technique of modifying filler surface properties in rubber compounds resulting in higher durability of the products. Even dispersion also suffers because of Van der Waals pressures which prefer aggregation and this could be a limitation to their efficacy (Fig. 2.1). As a result, several surface modification and functionalization methods have been invented in a way that encourages the dispersion of CNTs in polymers, thus, capitalizing on the synergistic benefits of its composite with other materials to the greatest extent possible [3].

Thus, there have been twin solutions to this problem discovered. The former solution is observed to be both realistic and stable whereas the latter solution is not stable when the least eigenvalue technique is used. The obtained results are also compared to the already published ones in order to prove the accuracy of the results and it depicts that the given work is accurate up to the fourth decimal place. In addition, the streamline profiles are also shown alongside the minimal eigenvalue profiles so as to explain the flow patterns. The position of the flow separation is found. Lastly, in addition to, mechanical and electrical advantages, carbon nanotubes improve the thermal stability of composites. CNTs also allow improving the thermal conductivity of polymers and increasing their resistance to deformation and degradation under the action of heat during processing and over the entire service life [4].

The energy problem, along with the need to develop the most suitable technology, which would be applied in the storing option of the energy, has found the rechargeable zinc -air batteries (RZABs) as a safe battery having high energy density. HEPBAs also templatizing, a specialized process of producing complex structures, of course. The nanoparticles prepared using HEPBA (HEPBA/CNT-800) are more active than nanoparticles prepared using HEPBA (HEPBA/CNT-800) regarding the performance on ORR ($E_{1/2} = 0.77$ V) in 0.1M KOH solution and OER (50 mA $cm^2 = 330$ mV) in 1M KOH solution. The RZAB with HEPBA/CNT air air electrode displayed an open-circuit voltage of 1.39 V with an exceptional energy density surface of 71 mW cm^2. In addition, the charge and discharge cycles of up to 40 h at 5 mA cm^2 current density characterize its high stability. In this paper, a new direction to systematically creating HEPBA derivatives has been provided to make a sustainable rechargeable metal Air battery system. It also displays the recent trend that due to improved thermal stability of the composites in the incorporation of the CNTs, there is a possibility of developing high life and reliability to withstand hostile environments [5].

To conclude, carbon nanotubes are multifunctional reinforcements in composite materials, as they enhance their mechanical strength, electrical conductivity, thermal stability and the general durability. Their dispersion and surface functionalization abilities open a fresh chance of material breakthrough in a variety of fields, including nanomedicine, aerospace, and electronics. Further investigation is needed to fully utilize the carbon nanotubes potential, possibly leading to a great advance in material science and new applications.

2.1 Introduction

SWCNTs are nanomaterials that have an enviable number of properties. The objective of SWCNTs chemical modification is most often the tuning of their intrinsic properties. Most recently, the mechanical bond has also been incorporated to the chemical toolbox to functionalize SWNTs. One starts this Tutorial Review by asking what is attractive about the mechanical bond as pertains to materials science in general and single-walled carbon nanotubes in particular. We proceed to describe the various synthesis routes of single-walled carbon nanotubes, emphasizing on their similarities as a prospective future synthesis route map. and lastly, we demonstrate with examples how single-walled carbon nanotubes synthesis can alter the surface property of SWCNTs, alter their electrical behavior and bind them with functional molecular fragments. This Review has the following goals: (i) to bring the reader up to date on the current practice of mechanical bond chemistry with SWCNTs: under what conditions it can be used, how it is actually carried out in practice, what input it has already made towards applications, and what it is capable of achieving in the future. The other important technical aspect of carbon nanotubes is extraordinary mechanical strength, i.e., it is sometimes stated to be some 100 times stronger than steel, and lighter than aluminum [6].

The greater strength to weight ratio renders the multi-walled carbon nanotubes very useful in polymer composites whereby mechanically reinforcing is of concern. In the last one decade, Nanomedicine research has achieved immense strides in the diagnosis, monitoring and treatment of retinal diseases. More recent reports indicate that nanomedicines may consist of the next generation of therapeutic agent in the neovascular complications, extend the half-life of therapeutic agents, serve as an alternative to sustained drug delivery vehicle and serve as scaffolds in retinal tissue engineering. Nanomedicine has a significant potential to transform the paradigms of the existing form of treatment by overcoming the shortcoming of invasive treatment, performing targeted drug delivery, creating sustained-release forms and stimulating tissue regeneration in treating retinal diseases [7].

Concerning electrical characteristics, carbon nanotubes possess outstanding conductivity which can be highly improved upon in composite materials. The nanoparticles of carbon present the prospective dangers to the growth and stress resistance of the plants. One of these carbon nanoparticles, fullerol (polyhydroxy fullerene), is able to enhance biomass generation in several drought-stressed plants but it is not known at the molecular and metabolic level how fullerol ameliorates drought stress in Brassica napus. In current experiment, exogenous fullerol was applied to leaves of B. Brassica napus seedlings exposed to drought stress. Further, regarding the secondary metabolite's metabolism, B. Brassica napus under the soil desiccation stress condition seen the decrease in the quantity of the phenolic compounds and flavonoids, including hyperoside and trans-3-coumaric acid. But, the accumulation of carbs in B treated with fullerol did not change significantly. Drought-stressed Napus. This is supplemented by their quantum peculiarities that make possible efficient charges transfer inside the material [8].

The total conductivity can also be enhanced by strong interfacial forces between carbon nanotubes and polymer matrices that tend to reduce the electron scattering. Flexible sensing technology has many applications in the meaning of smart production and artificial intelligence of robots. Sensitivity is the important parameter in the consideration of the sensors. Nevertheless, the investigations of the impact of the geometric size of porous flexible sensors on sensitivity are not full yet. Our study indicates that, the sensitivity can be enhanced by reducing the thickness of dielectric layer and enlarging the size of the electrode. They have discovered that CNT reinforced composites may surpass the conventional materials in electrical conductivity properties making them specifically useful in the area of developing sensors where charge transfer speed is of great concern [9].

In this paper, a technological foundation of next-generation power grid transformer (NGPGT) to interlace renewable energy is created. The new energy systems have focused on the renewable energy sources productions, transmission and distributions whereby the interaction with the grid transformer is contributory to about 70 percent in the processes. The use of the present-day framework power grid transformers in terms of renewable energy integration could be limited by absence of centralization, local control, interoperability, genuineness and proper bi-directional flow. The present paper has been able to invent a new technological idea, that is, NGPGT which is more functional to overcome the deficiencies of the transformers of the present generation, NGPGT. This is set up with the involvement of high-tech flexibility to the existing system that constitutes automatic condition monitoring, smarter inverters, edge computing, automatic control, and smart management. The benefits, together with the coverage of the NGPGT in relation to the existence of current transformer, has been reflected in this research paper through the summation of the most outstanding specifications and the components that are needed. It is also in this report that the existence of some of the problems to deal with the implementation of NGPGT in regards to operation, communication, energy management and economic view point, consequently providing the scholars with an opportunity to conduct more studies. Enhancement of thermo-conductivity of polymer composites with carbon nanotubes can be applied in industries that temperature control is of particular concern, i.e. electronics and car manufacturing [10].

Tangential flow interfacial self-assembly (TaFISA) can be conceivably as a Scaleable process because it can be used to orient the uniform arrangements of carbon nanotubes towards high performance semiconductor circuits. Assembly at an interface between liquid and liquid and subsequently attached to the patterned substrates. Having used a substrate with a different chemical functionalization we have ensured that well aligned, high density, individual nanotubes would only be capable of deposition at somewhat low water contact angles between 35 and 65. Water contact angles (WCA) less than ten degrees lead to the nanotubes aggregating thus leading to the non-uniform delivery of the film and at high water contact angles above sixty five degrees the align deposition procedure is stopped. The most alignment is observed to be achieved by the ratio of low wrapper polymer to nanotube, 0.6:1, and with the least defect of liquid crystals defects when the polymer wrapped nanotubes are dispersed in chloroform. We demonstrate that with the alteration of the water subphase to include glycerol we are able to increase the global orientation and decrease the imperfections of the liquid crystals and also increase local clumps of nanotubes. The observations are useful in showing the way to go when achieving TaFISA as well as applying it towards the production of technologies that would be compatible with the semiconducting carbon nanotubes, with the adjusted natures. Carbon nanotubes the electrical band structure of the carbon nanotubes may be tuned relative to diameter and chirality. These considerations play an important role regarding the primarily metallic or semiconducting properties of a carbon nanotube [11]. The increase in the demand of the high-performance electrons on a low-cost basis has boosted the innovation of new high-performance dielectric materials. We were also already past the research on how the operating frequency of the device affects the performances and had come to conclusion on the fact that when the operating frequency of the TFTs was lowered the hysteresis was reduced. As it can be observed, a greater reduction in hysteresis was achieved in 35 mol% polymer. Different electrical characteristics of carbon nanotubes synthesized with CVD are with helical structures: they offer different channel directions of charge transport inside the nanotube structure [12].

They are made to under non-covalent modifications in order to make sensors that Fluoresce basing on biomolecule binding. Non covalent chemistry however, has many disadvantages that makes reliable identification of molecules and signal transduction impossible. We have developed a general covalent strategy to the manufacture of molecular sensors which does not affect fluorescence in the near-infrared (>1000nm) in a deleterious manner. Using this method, it is possible to integrate more recognition units therefore allowing the universal approach to NIR fluorescent biosensors with improved extents of stability. The most recent developments have revealed what carbon nanotubes (CNTs) can do in the various optoelectronic applications whereby their extraordinary photo conductive properties and direct emission have enabled the insertion of CNTs in display technology and biosensors [13].

2.2 Electrical Properties of Carbon Nanotubes

The remarkable electrical properties of CNTs have attracted much interest in the recent past, properties that have been credited to the unusual one-dimensional geometric structure, and the complex interdependence between geometrical structure and electronic properties. The future of carbon-based electronics is entrusted on mass production of high purity high yield semiconducting carbon nanotubes. We have established new set of extraction process to isolate SWCNTs in better yields and purity separately. s-SWCNTs with large diameters are preferentially encapsulated by PCz and the sorting efficiency could be increased fourfold by absorption and Ram After mentioning that the ordered s-SWCNTs could be aligned into field-effect transistors and photodetectors both of these applications demonstrated unusually semiconductor characteristics and high spectrum detectivity in combination with satisfactory long-term stability. The other prominent characteristic of carbon nanotubes is that they might be either metallic or semiconducting and this propensity is mainly determined by the chirality, diameter, and the surface chemistry of the tube [14].

The Hummers process was used to synthesize graphene oxide/sulfur combination. The samples were evaluated in ambient temperature in air by measuring the current voltage parameters. The I-V characteristic curve of the graphene oxide/sulphur mixture, having oxygen to carbon ratio of 3.54 and sulphur to carbon ratio of 42.54, displayed negative differential resistance (Fig. 2.2). The transitions of current carriers between the localized states formed by functional groups are attributed to the negative differential resistance [15].

In addition, the charge transport phenomena of carbon nanotubes are also quite appealing where the large mobility and the mean free paths are the source of the main factors that contribute to the high-performance electronic device. Transition metal dichalcogenides

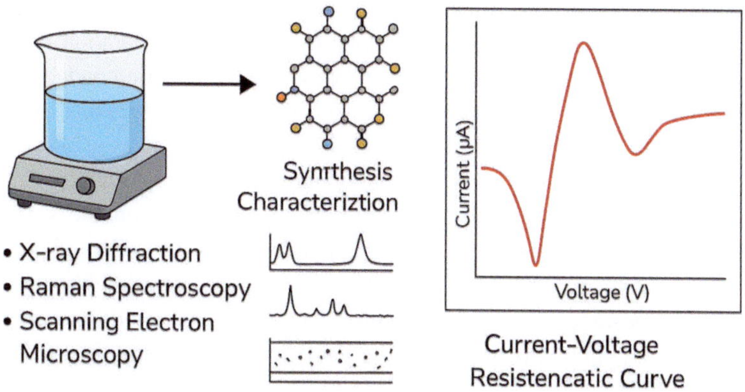

Fig. 2.2 Negative differential resistance of graphene oxide/sulfur compound

(TMDs) monolayers present a canonical two-dimensional system to uncover a wide variety of electrical properties and potential applications with their chemical diversity. TMD nanotubes (SW-TMDNTs), belong to the 1D materials, with rolled TMD monolayers at the form of defect-free cylinders, which electrical properties may be engineered through chirality and composition. Much less however has been done to study their geometrical arrangements as well as their chemical reactions owing to their inability to be stable ordinarily. The process is also applicable in the preparation of other TMDNTs including selenides and alloyed $Mo_{1x}W_xS_2$. Broad-reaching microscopic and spectroscopic experiments allude to the illustrative assemblage of the Janus $MoS_2(10)Se_{2x}$ nanotubes [16].

Their quasi-one-dimensional nature brings down the number of scattering events, thus whenever a high efficiency in the transport of electrons is required, as in the implementation of high-speed electronic systems, it is favored. A wide review of structure-property correlation of carbon nanotubes demonstrated that a change in the synthesis approach may enhance the electrical performance either by a post synthesis modification or by composite configurations where CNTs are inserted into diverse matrices [17].

As is described in this paper, sulfur doped graphite oxide/Graphene oxide sheets constitute a probable functional material, being quite simple to generate the sheets in a one-pot, single batch, composite. This allows the existing voltage properties to demonstrate the negative effect of the differential resistance (NDR) due to an effective composition of the properties constituents of the materials on the basis of the structures as well as the electrical properties whose status is the three-dimensional form of an interlaced graphene oxide sheet particles grid placed inside the stacking layers of the sheet of the graphene oxide that the colleagues are connected to one another. The conducting channels of the proton hopping caused by the adsorption of water molecules on the hydrophilic layers of graphene are, as a matter of fact, not stable due to the transport thermal movement of the molecules, which will lead to emptying of the conducting channels and the negative variation of the resistance (NDR). It is anticipated that the incorporation of the sulphur atoms into graphite oxide lattice can lead to enhancing the electric conductivity and the NDR performance owing to the increased transfer of electrons to graphite oxide particles and increased lifetime of the H + transport ions [18].

The photophysics character of CNTs are undoubtedly intriguing. Photosensitivity spectrum of carbon nanotubes (CNTs) was investigated depending on the values of bias voltage in the range of 400–900 nm, and with five values U = 1 V, 3 V, 5 V, 7 V and 9 V. The research paper has revealed that bias voltage elevation on the CNT sample causes the appearance of a greater range of negligible or low photosensitivity and the reduction of the photosensitivity peak heights. Specifically, it is seen that instead of being photosensitive at U = 9 V, the sample has become irrelevant across most of the wavelengths band (400–900 nm) and is therefore no use as a detecting device with the mentioned value of voltage. The greatest photosensitivity range, as well as the photocurrent maximum, are recorded at U = 1 V voltage, and at the wavelength of 715 nm the most significant photocurrent peak is obtained, which is equal to Ip = 0.2 uA. It means that carbon nanotubes with voltage of U = 1 V can be successfully used as radiation detectors with a broad range of radiations. The experimental values of I-V characteristics of the investigated nanotube samples are obtained in six wavelengths, 640, 720, 740, 760, 780 and 880 nm in points and curves. The approximation curves are exactly described as the fourth-degree polynomials. It has been revealed that the I-V characteristic at 2 has a small photosensitivity in the almost whole range of the voltages extending to 1–9 V as at 720 nm. This measurement is achieved with a longer wavelength; this is because the photosensitivity peak is projected to fall to lower voltages, but the photocurrent peak declines with any wavelength [19].

The article has examined photoconductive behavior of multiwall carbon nanotubes. They investigated photosensitivity spectrum dependence of gadolinium-doped multiwall carbon nanotubes (10% Gd). Spectral absorption of the gadolinium - doped MWCNTs has been measured between the wavelength of 400 nm to 900 nm, and with a voltage of 1 to 9 V. A photocurrent peak of about 0.158micro A was obtained at a wavelength of 480nm with 3V applied voltage, this is the highest sensitivity. The other maximum intensity is gained at the wavelength of about 540 nm. with the application of U = 5V, photosensitivity is lost and sacrificed. The optimum range of photosensitivity occurs at U = 1 V. Application of this value of voltage might be useful in detecting applications. They provide an experimental data of six wavelengths within spectral range 640–880 nm and show presented I-V characteristics. The I-V curve demonstrated low photosensitivity with the whole range of voltages of 1 to 9 V under l-640 nm. The values of the photocurrent have the tendency to increase because of the increasing wavelengths of 720, 760, 780 and 880 nm.

The maximum photosensitivity is observed at the voltage of 7 V. The results suggest a positive future of the cost-effective and scalable synthesis of Gd-doped MWCNTs to apply in detectors and optoelectronic devices [43]. Photoelectrical properties of carboxyl-functionalized carbon nanotubes (fCNTs) were studied by their synthesis. Photosensitivity spectra of a carboxyl-functionalized CNT sample were measured at a voltage of between 1 and 9 V and wavelength of between 400 and 900 nm. An applied voltage of 1 V resulted in lowered photosensitivity across the wide band wavelength (visible to near-infrared). fCNTs showed the optimum photocurrents at 5 V at wavelength range λp ~ 400–800 nm and at 3 V at the wide spectral range λp ~ 400–900 nm. Experimental data were analyzed which allowed determining the widest range of photosensitivity and the peak sensitivity value. Therefore, an output voltage of U = 5V was obtained. The highest photocurrent peak Ip 2.67-m was observed at the wavelength of 720 nm. The photosensitivity spectrum of fCNTs and pure CNTs suggest that the photosensitivity of the former is significantly enhanced. The fCNTs have the peak photosensitivity of Ip ~ 2.67 μA, whereas pure CNTs have Ip ~ 0.185 0. It has measured up to 14-fold enhancement of photosensitivity of fCNT. The mathematical analysis of spectrum dependencies of the produced photocurrents with applied voltages of a variable voltage may be expressed in terms of fourth order polynomials (Fig. 2.3). The corresponding I-V characteristics at two wavelengths of 760 nm and 780 nm follow the same tendency with the peak photocurrent being located at lower voltages. One of the most practical applications of carboxyl-functionalized nanotubes is the light detector and optoelectronic applications [20].

The volt-ampere characteristics curve at oxygen to carbon ratio of 3.54 and a grey ratio of 42.54 showed negative differential resistance. Current–voltage characteristic part modelling (including the negative differential conductivity portion) of two samples of structures composed of graphene oxide/sulfur compounds and the entire volt-ampere characteristic curve was carried out. It is depicted that the regions of negative differential conductivity of the first type of samples can be well described by the third and

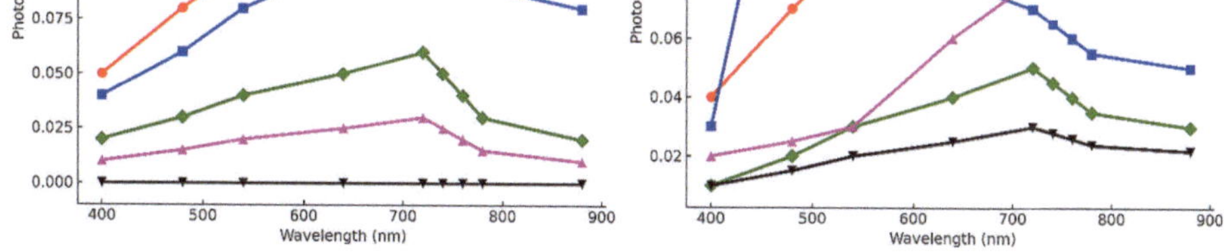

Fig. 2.3 Photosensitivity spectrum of pure CNTs (**a**) and gadolinium doped CNTs

fourth-order polynomials, and the second type of samples by a fifth-order polynomial. The second type of samples had volt-ampere characteristics within the range of 6–17 V that can be described by a linear model, but the characteristics within the range of 21–32 V follow a second-order polynomial model. The maximum error of approximation curve is at U = 26 V. These structures voltampere characteristics over the entire voltage range are well modeled by both a power function and a second-order polynomial [21].

Carbon nanotubes in general have extraordinary electrical properties owing to their one-dimensional (1D) electronic structure and quantum size effects. A carbon nanotube (CNT) has electrical properties that are highly dependent on its chirality, defined by indices (n,m), and define according to the electronic type of the nanotube.

Chirality and Bandgap:

- In the case of SWCNTs, the condition n–m = 3q with q an integer, leads to metallic behavior.
- The CNT on the other hand is a semiconductor with an energy bandgap Eg.

Bandgap equation for semiconducting single-walled carbon nanotubes (SWCNTs):

$$E_g = \frac{2\gamma_0 \alpha}{d_t}$$

where:

$\gamma_0 \approx 2,9eV$—nearest-neighbor hopping energy;

$\alpha = 0.246nm$—graphene lattice constant;

d_t—nanotube diameter.

Metallic carbon nanotubes can exhibit ballistic transport of electrons over micron distance due to minimal scattering.

$$G = \frac{2e^2}{h} M$$

where:

G—conductance;

E—electron charge;

H—Planck's constant;

M—number of conduction channels.

This relates to a quantum conductance measurement of:

$$G_0 = \frac{2e^2}{h} \approx 77,5\mu S$$

Carbon nanotubes have exceptional current density capabilities, surpassing $10^9 \frac{A}{cm^2}$, making them appropriate for nanoelectronics interconnects.

A second major area of CNT research is their interaction with biological Systems and materials to use them in bioengineering. Carbon nanotubes have demonstrated great promise in biosensors in the detection of biological materials due to the large surface area and electrical sensitivity. Studies in the area of carbon nanotubes (CNTs) sensor design have demonstrated that their outstanding electrical properties can be enhanced through covalent functionalization that increases biocompatibility and sensor selectivity [22, 23]. These successes manifest the versatility of CNTs, not only as electrical devices, but also as a way to initiate new methodologies in biochemical sensing, and thus expanding the health care diagnostics.

In short, carbon nanotubes present extraordinary electric properties because of their diversity in structures and flexibility. Their applicability on electronic gadgets, biosensors as well as optoelectronic systems marks a distinguished advancement of nanotechnology and materials science. Further research towards optimization, synthesis and integration of carbon nanotubes with other materials will be important towards their absorption into future technologies to cement their place as important components in future electronic circuits and devices.

2.3 Mechanical Properties of Carbon Nanotubes

The mechanical behavior of CNTs has drawn a lot of attention in the materials research due to their extraordinary nature surpassing the behavior of most of the conventional materials. Made of one-dimensional (1D) nanostructures of graphite, CNTs are exceptionally strong, stiff, and elastic, which makes them of necessity in the further development of a wide range of composite materials. It is acknowledged that the use of aerosol catalyst in flame synthesis of carbon nanotubes (CNT) results in the production of SWCNT, useful in a number of applications. There is a lack of modelling work to put the operational parameters of SWCNT development to the optimum. A model of the aerosol-catalyst system in flames is also baseline and the effect of oxygen on the growth of SWCNT is studied. Computational Fluid Dynamics simulation is utilized to form a baseline flame model of a standard diffusion flame with 24% oxygen content at the input. The entrainment catalyst particles are simulated by a dispersed phase model (DPM). The flame model is coupled with a known model of CNT growth rate to predict the growth rate of CNT at every particle. The oxygen concentration of the inlet is varied between 19–27 percent to study its effects on the development of SWCNT. Baseline flame shape and temperature have satisfactorily been proven valid. The results also show that the longest length of CNT is produced by particle 3 in the baseline scenario compared to other particles because particle 3 has the optimum synthesis pathway. The particles are classified based on their residence times namely shortest, intermediate and longest. An increase in oxygen contents of 19 percentage to 27 percentage results in a Lesser CNT length of particle 3 by 30 percentage under all input conditions due to the reduced carbon precursor and composition within the flame. It was found that regardless of the burner operating conditions, significant SWCNT growth is always expected in 120–140 mm HAB, which implies that there is an optimum range of species concentration, in which the SWCNT growth is favored in the aerosol-based flame synthesis. Subsequently, a conclusion can be made that the growth of SWCNTs through the aerosol-based method is greatly affected by the carbon source and slightly affected by temperature. Carbon nanotubes (CNTs) have the potential tensile strength that is 50 times higher than that of steel, yet they have a lightweight character with their Young modulus estimated to be approximately 1 TPa [24]. Such an extraordinary combination of properties holds great potential to be used in practice, in particular, in strengthening polymers and other materials.

The important mechanical properties of carbon nanotubes can be attributed to their atomic structure, unique carbon atoms are placed in a hexagonal lattice, as in graphene. It has an excellent tensile strength and high flexibility which enables it to bear enormous deformative loads without a structural collapse [25]. The studies state that the reinforcing effect of carbon nanotubes (CNTs) added to the polymer matrices has a significant positive impact on the mechanical property of the composite material and, thus, enhances the fracture toughness and fatigue resistance [26, 27].

One potential material of flexible and wearable thermoelectric (TE) generators is carbon nanotube (CNT) yarn due to its structural flexibility and high electrical conductivity. One of the main obstacles however in the progress is the control of n-type doping and also atmospheric stability which plays a critical role in the preparation of p-n junction pi-type cells and also the optimization of the same. An undesirable cause of doping which could pose a challenge towards controlling doping is one occasioned by the oxygen in the atmosphere. These findings indicate that polyoxyethylene (50) stearyl ether plays many different roles in the production of n-type CNT yarn: it acts as a surfactant to stabilize the CNTs in water, as an electron donor to render the CNTs n-type and as a protective cap to avoid the oxidative destruction of the CNT (Fig. 2.4). Also, it has been concurrently elucidated that low concentration of CNTs can result in a significant increase in the mechanical properties of the formulated composites, which underlines their appropriateness as a reinforcing agent [28].

As highlighted by many studies, the mechanical behavior of CNT-reinforced composites is strongly influenced by various factors among them being the dispersion state of CNTs in the composite matrix, the interfacial bonding between the CNTs and matrix as well as the alignment of the CNTs during processing [29, 30]. This is because the precise alignment and placement of the nanotubes are significant in maximizing the transfer of the loads between the carbon nanotubes and the matrix which has a direct impact on the overall mechanical performance of the composite [24, 31].

The mechanical performance of carbon nanotubes is superb owing to the strong sp 2 covalent bond within the graphene structure.

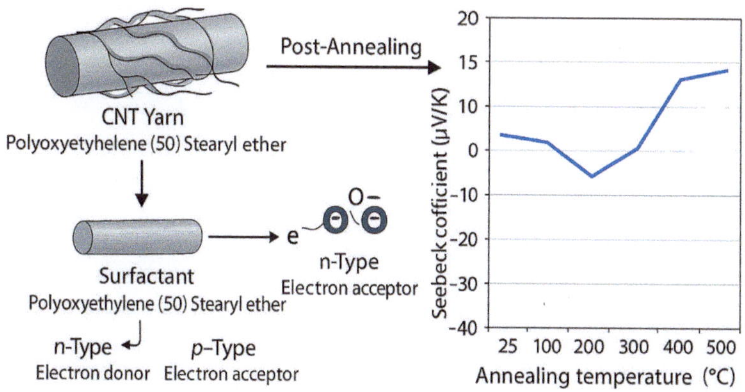

Fig. 2.4 Fabrication and stability of n-Type CNT Yarn via surfactant-assisted doping and thermal annealing

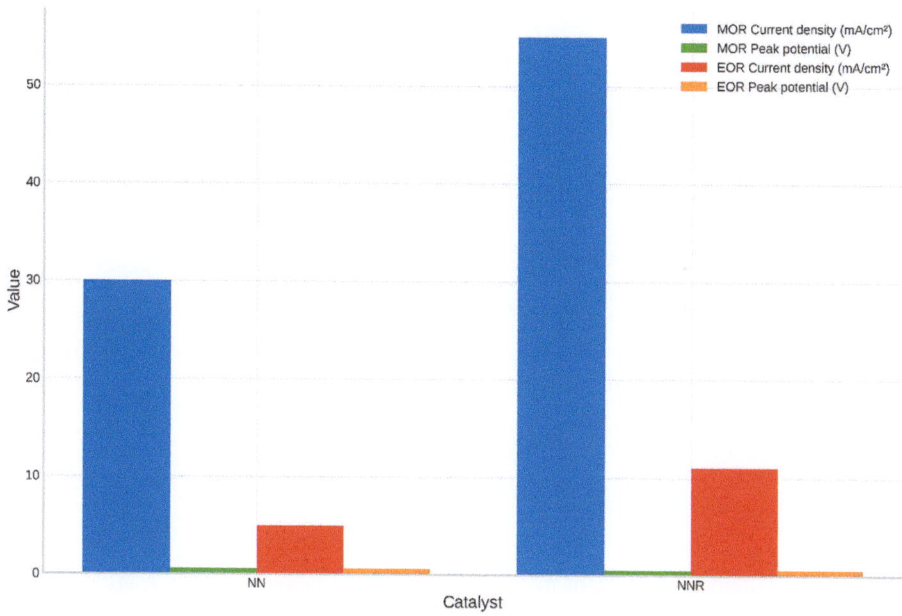

Fig. 2.5 Electrocatalytic performance of NN and NNR in MOR and EOR

This gives CNTs:

- Extremely high Young's modulus
- High tensile strength
- Remarkable flexibility and resilience.

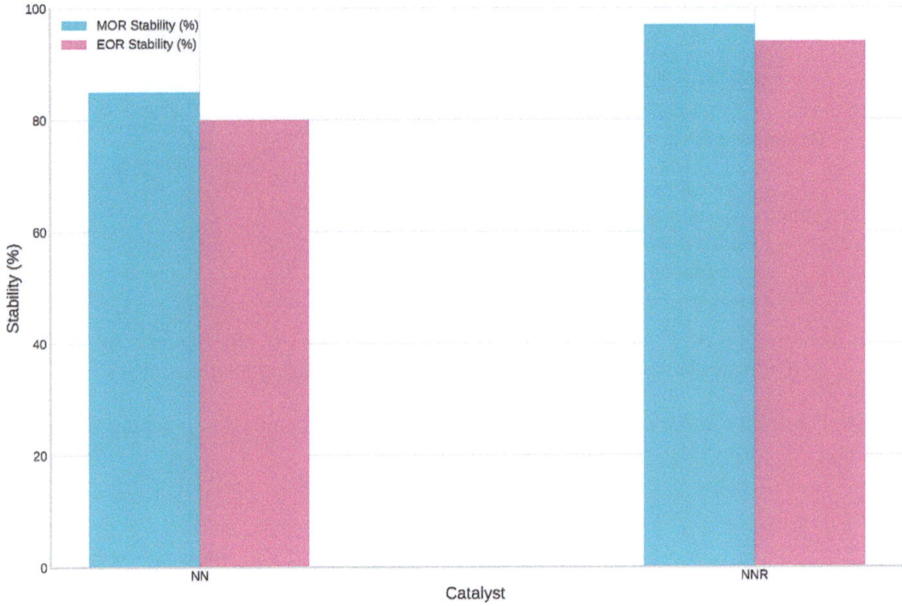

Fig. 2.6 Stability of NN and NNR after 1000 cycles

Young's Modulus:

For both SWCNTs and MWCNTs:

$$E \approx 1.0TPa$$

Tensile Strength:

$$\sigma_{tensile} \approx 50 - 150GPa$$

Flexibility:

The carbon nanotubes, because of their one-dimensional tubular geometry, are capable of sustaining large elastic deformations without causing permanent structural damages.

Continuum model for bending stiffness:

$$D = \frac{Eh^3}{12\left(1 - v^2\right)}$$

where:

- h—wall thickness (typically taken as ~ 0.34 nm)
- ν—Poisson's ratio (~0.19 for graphene).

Carbon nanotubes possess very high strengths-to-weight ratios, and therefore some measures of their strengths are higher than many known engineering materials. We describe a single-step hydrothermal synthesis of hybrid nickel sulphides, i.e. Ni_3S_4-NiS (NN) and Ni_3S_4-NiS-rGO (NNR), with reduced graphene oxide (rGO) to use as catalysts (Figs. 2.5, 2.6). CNTs in conduction polymers have demonstrated improvement in electrical conductivity and mechanical reinforcement making them multifunctional such that they are electrically conductive and mechanically strong [32].

The research proves that alumina nanofibers addition makes the coating harder, more flexible, and resistant to corrosion. The functionalization that occurs due to the completion of alumina nanofibers in a silane-toluene solution leads to better dispersion the usage of nanofibers into the polymeric structure of the coating, and thus better performance properties (Fig. 2.7).

Carbon nanotubes flexibility is reflected in the ability to undergo various modifications to tailor some of their properties and characteristics, including chemical functionalization or hybridization with other nanomaterials [33]. Such modifications can enhance the interfacial behaviors in composites leading to better mechanical behaviors. Of importance are the nature of the surface states of the nano dispersed filler particles. The issue of dispersing the nanomaterial in a polymer matrix could not offer an effective solution unless the surface nature of the nanomaterial filler is matched with that of the polymer matrix in the composite materials so formed. The environmental sustainability consideration in addition to the kinetical properties led to the study on CNTs in bio composites and green

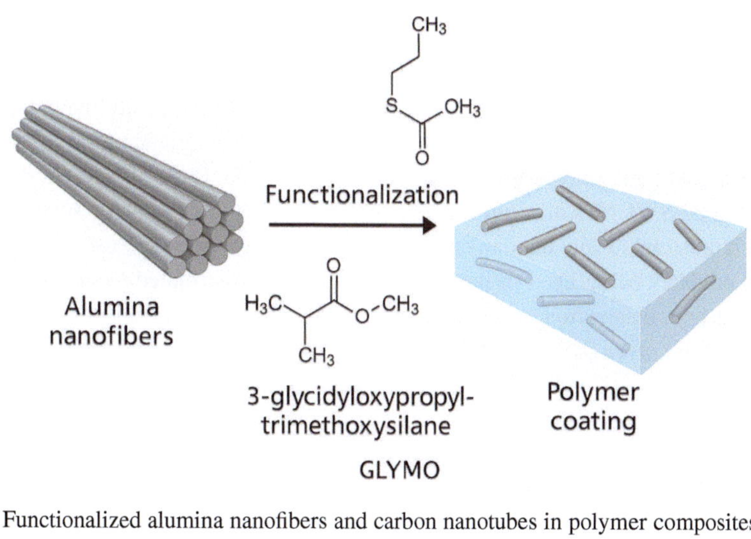

Fig. 2.7 Functionalized alumina nanofibers and carbon nanotubes in polymer composites

composites. Combinations of carbon nanotubes with bio-based matrices could result in highly strong and environmentally friendly materials, which is beneficial in such industries as automotive and construction, where reducing the impact on the environment is increasingly important [34]. The researchers, who are working on the greener manufacturing processes and bio-based raw materials, consider CNTs to be the key to developing the sustainable high-performance materials. Besides, advanced CNTs applications continue to be discovered mainly in the fields of nanotechnology and biomaterials. Their extraordinary mechanical properties pave the way to new achievements in creating flexible electronics, medical devices, drug delivery systems, which require materials that may resist mechanical stresses but allow promoting biological interactions [35, 36].

Carbon nanotubes represent a remarkable new family of materials whose mechanical properties are very superior to those of conventional materials. Their inclusion into various composites enables the progress in multiple industries. The body of literature informs about the necessity of the additional research on CNTs to achieve their full potential and address the concerns regarding the scalability of the manufacturing process and environmental effects. Finally, the mechanical properties of carbon nanotubes including extraordinary strength, stiffness and flexibility are enabling new applications in novel materials and they will be inevitable in the further technological developments.

2.3.1 Thermal Properties of Carbon Nanotubes

The unique nanostructure with very high thermal conductivity has made heat properties of CNTs to be the topic of numerous studies in the recent years. There is a distinguished acceptance of two primary categories of carbon nanotubes which include the SWCNTs and MWCNTs and they are both different which pertains to the property of thermal conductivity.

An appreciation of the manner in which these characteristics become influenced by the structural factors is vital to various applications namely composite materials and heat management systems in electronic applications. The second significant property of CNTs is their excellent thermal conductivity that can be as high as 1500 W/(mK) in the example of well aligned SWCNTs at room temperature. An exceptional value of such character is sharply contrasting to the standard materials that would ensure the use of CNTs as the perfect candidates being used in the domains that would be related to the need to generate efficient heat dissipation and heat management. The thermal transport characteristics of the CNTs are strongly affected by a number of significant facts, of which structural orientation, cleanliness and defects are associated. Quite often, e.g., with CNT structures being out of alignment, the thermal conductivity may be considerably diminished and reports as low as below 0.1 W/(mK) have been transcribed. The architecture of CNTs as a hierarchy affects the heat conductivity in it rather significantly. The thermal transport properties of MWCNTs are found enhanced by the interaction between the various walls than those between an individual shell to yield a net thermal conductivity exceeding the sum of the individual conductivities [37, 38].

The heat conductivity of these substances is promoted with the help of CNTs and therefore they can be applied to the aerospace industry and automobiles. The foreseen gap in the existing research that may be filled with the introduction of small percentages of MWCNT, TiN, and c-BN powder into clear resin is the topic of the new perspective. According to the research, the hardness and improved thermal properties are clearly observed and consequently the importance and novelty of research in such enhancement of composite materials. In the present study, it is also demonstrated that there exist significant changes in the hardness, including thermal conductivity due to the use of nanofillers. It depicted the effectiveness of c-BN to improving mechanical properties since the hardness had a 20.45 increase of 88.53 to 106.64 Shore D upon adding 1 wt.% c-BN. Conversely, adding 1 percent of TiN to the present composites attained hardness of 80.76 Shore D, a clear representation that it is capable of decreasing hardness of composites. Besides, the strengthening 1 percent MWCNT resulted in 24.21 percent boosting of thermal conductivity to enhance thermal transport properties of CNTs coefficient (0.19 to 2.36 W/m k). SEM analysis phase showed that TiN curing was a problem due to its density and color and the layer thickness of 0.02 mm and less should be used to favor curing. The viscosity of the pre-production mixture has an informed relationship with the final hardness of the product which proves the overall effect it has towards the final mechanical properties. Additional investigations have to be conducted to study the inclusion of other forms of nanofillers, such as carbon nanotubes (CNT) and boron nitride (BN). Exploration of hybrid fillers as e.g. CNT + BN or combinations of other fillers might provide even better performance benefits than mono component fillers at same filler dose may be of benefit. Under such tests so much superior kinetical properties can be disclosed. An extremely impressive alteration in the improvement of thermal bearing and the deterioration of features in carbon nanotubes in SLA three-dimensional practical resin-based composite has also been made that has significance beyond measure in the elements when presented to extreme temperature during application [39], (Fig. 2.8).

This paper has synthesized CNT and FCNT. The CNTs and FCNTs are cylinder shaped and the surfaces are almost smooth and uniform in character. Chemical conversion and thermal property of the synthesized CNTs and FCNTs were assured by the TGA measurements. In high temperatures above 5000C carbon nanotubes are destroyed in a characteristic fashion due to their thermal instability. The intermolecular forces among carboxylic group (9 percent reduction of FI to 500 °C in FCNTs) and vaporization of physically adsorbed body of water molecules trigger the reduction of mass. The samples too improved the specific heat capacity Cp at constant temperature with an increase in temperature. At the specific temperature T = 1086.3 K and T = 916.3 K, the specific heat capacity at the certain temperature, Cp, of the carbon nanotubes (CNTs) and functionalized carbon nanotubes (FCNTs) is maximal respectively [40], (Fig. 2.9).

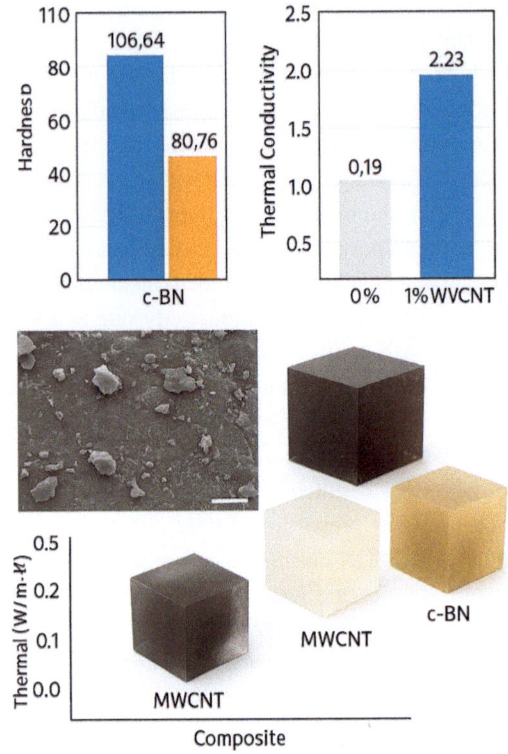

Fig. 2.8 Performance of nano-filler reinforced transparent resin composites

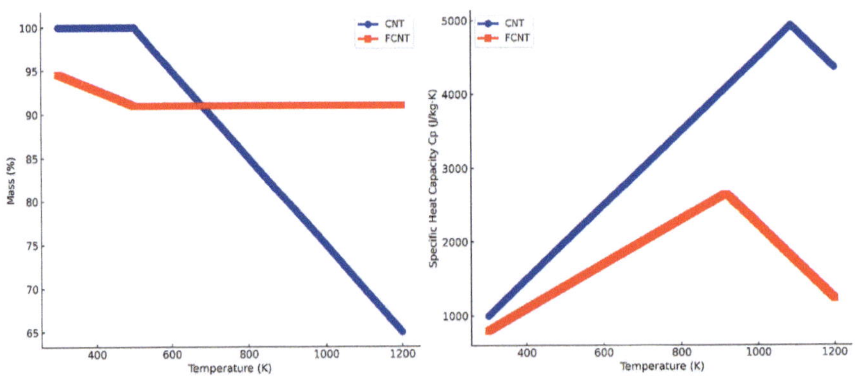

Fig. 2.9 Thermal analysis of carbon nanotubes (CNTs) and functionalized carbon nanotubes (FCNTs)

The process of synthesis of graphene oxide through Hummers method gives Sulphur concentration that has yet been covered in little publications. Under ambient condition we produced thermally stable multilayer sulfur-doped graphene oxide by the aid of optimized Hammers method. The samples subjected to experimentations were pre-heated to a predetermined level in the electric-furnace at varied ambient temperature to determine the impact of the thermal condition on the structural and thermo-physical state of the samples. In this process correct measure was obtained to find out the difference in crystallite dimension and linear coefficient of thermal expansion (CTE). These parameters were obtained by use of the measures of variation of the ratio of intensities of the characteristic Raman spectral bands with rising of temperature. The shift in Raman peak intensity also revealed temperature dependence which seemed to be of utility in assessing microstructural development as well as residual thermal activity of material during the heat treatment cycle. Under the high temperature condition, less dimensional shrinking was reported to be caused using multilayer graphene oxide doped at 2 wt% sulfur (carbon to oxygen ratio of 2.3), compared to undoped graphene oxide. This shrinkage was on an extremely narrow temperature gap between 308 318 K and resulted in extremely low coefficient of linearity in thermal expansion, i.e. -18 ppmK (Fig. 2.9). The achieved results of the experiments are quite good in the framework of the real area of different advanced technologies, at least those mentioned solar harvesting systems, sensing, solid lubricants, thermally active actuators, and materials that can be made to wave like the transport phenomenon phenomena such as second sound [41].

The paper discusses the preparation and analysis of the thermal properties of 10 and 15 percent gadolinium doped MWCNT. The TEM methods were employed to investigate the microstructure. It was shown upon morphological study of gadolinium-doped multi-walled carbon nanotubes that the surfaces were even and rather smooth, a characteristic of excellent structural integrity and homogeneity of the doping. Such consistency in the morphology of the surfaces played a major role in determining the general physical characteristics of the nanotubes. Successful and uniform doping was further confirmed by the even distribution of gadolinium on MWCNT surface through compositional characterization. The thermal behavior of the synthesized Gd-MWCNTs comprising 10% and 15% gadolinium by weight was examined with the help of thermogravimetric analysis (TGA). The mass change on heating was explained by the thermal decomposition of the carbon based-Matrix and de-composition of the Gd related intermediate compounds synthesized. These heat occasions gave indications of the stability of the material at high temperatures. The specific heat capacity (Cp) of the doped samples was measured by the differential scanning calorimetry (DSC). The results have shown that 10% and 15% Gd-doped MWCNTs followed the similar trends towards the changing of Cp with temperature. The highest specific heat capacity of the sample consisting of 10 percent Gd was determined to be $C_p = 2.88815$ kJ/kg K at $T = 916.290$ K and that in the 15 percent Gd-doped sample was the highest at which $T = 1106.290$ K and $C_p = 1.10801$ kJ/kg K (Fig. 2.10). These

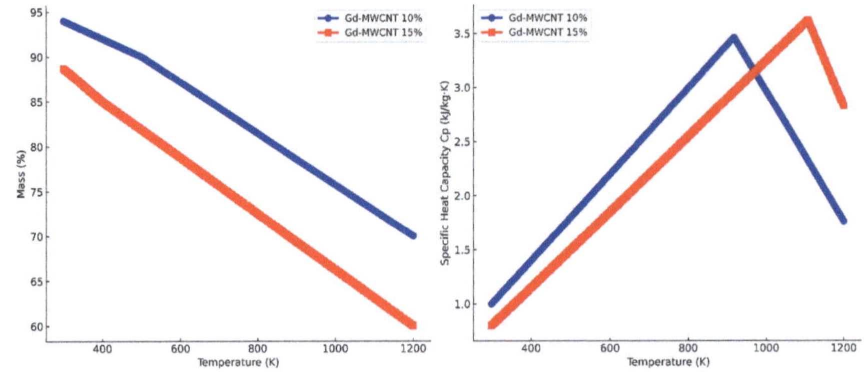

Fig. 2.10 Thermal analysis of gadolinium-doped multiwall carbon nanotubes (Gd-MWCNTs) with 10% and 15% Gd content

differences in the thermal capacities correspond to the effect of the dopant concentration on energy capacity and heat dissipation of the nanomaterial [42, 43].

The walnut shells used in this study provided porous carbon materials whose preparation was by a two-step thermal processing technique using carbonization at 800 °C, and activation at 400 °C after 1, 2 and 3 h respectively. Analysis showed that the activation time of a specific temperature results in a progressive decrease in the graphite crystallites with the decrease in the thickness and lateral size. Mean crystallite sizes were measured quantitatively, and they decreased, by a factor of nearly two, compared to thermally carbonized samples, by an amount of about 8 nm to around 5 nm, reflecting the refinement of the graphitic structures on activation. Furthermore, the thermos treatment at the temperatures of 400 °C resulted in the development of well-prepared mesoporous structure, the mesopore content of which varied between 75 and 78%, and the average mesopore width was approximately equal to 5 nm. This increase in porosity led to a provisional doubling of the specific surface area of the carbon material to that of the original, non-activated phase change material (PCM). The fractal dimension of the product porous carbon materials was also derived at low (P/P 0 < 0.4) and high (P/P 0 > 0.4) relative pressures with Frenkel Halsey Hill (FHH) method. The measurements gave more understanding on both surface roughness of carbon framework as well as pore connectivity and it confirmed the hierarchical porous structure formed during the activation process [44].

The carbon compounds so synthesized were made into electrodes of electrochemical capacitor with a 6M KOH electrolyte. The CVA technique, galvanostatic cycling and SEI were used in electrochemical behavior analysis of these systems and capacitance values specific to these systems were obtained. Electrochemical testing of the pre-treated carbon material (2 h, 400 °C) showed a specific capacitance of between 110–130 F/g against the discharge current density (4–100 mA/g). This was done by carrying out the assessment at a maximum charging voltage of 1V with Solid Electrolyte Interface (SEI) method. In addition, carbon nanotubes have an outstanding ability to augment the thermal management of electric equipment. The property highlights a recent tendency in utilizing carbon nanotubes and other carbon materials as fillers to enhance the thermophysical properties of mainstream materials to develop composites with specific thermal properties applications [45, 46].

Carbon nanotubes (CNTs) thermo-conductivity is one of the highest among all the known materials, which is explained mainly by:

- Strong sp^2 bonds → high phonon velocity;
- Low phonon–phonon scattering in defect-free tubes;
- One-dimensional phonon transport.

Thermal Conductivity:

For SWCNTs:

$$k_{SWCNT} \approx 3500 \frac{W}{m \cdot K}$$

For MWCNTs:

$$k_{MWCNT} \approx 3000 \frac{W}{m \cdot K}$$

Phonon Transport Equation:

Thermal conductivity k can be expressed as:

$$k = \frac{1}{3} C_v v_s l$$

where:

- C_V—phonon heat capacity;
- v_S—speed of sound (phonon velocity);
- l—phonon mean free path.

In high quality carbon nanotubes, the mean free path can be many micrometers long, and heat transfer can be almost ballistic.

Temperature Dependence:

At low temperatures, $k \propto T^3$ (due to boundary scattering).

At high temperatures, Umklapp scattering dominates, leading to $k \propto \frac{1}{T}$.

CNTs are used in:

- Thermal interface materials (TIMs);
- Heat spreaders in microelectronics;
- Thermal management coatings.

The thermoelectric parameter (zT) of the n-type Zintl compound $Mg_3(Sb,Bi)_2$ is excellent at wide temperature range, making it up to date with its large zT in a wide range of temperatures, yet its optimum performances have been optimized to room temperature range to be applied in practical operations of the advanced thermoelectric cooling and power generating systems. Evolution of the path towards enhanced thermoelectric performance of $Mg_3(Sb,Bi)_2$ by phonon spectrum mismatch is explored in this work and it is achieved by the use of multi-walled carbon nanotubes (MWCNTs). Even at moderate loading rates (as low as 0.5 wt%) the MWCNTs result in severe increase in the interfacial thermal resistance and the net effect is to inhibit the lattice thermal conductivity and provide an excellent thermal power factor. The composite containing 0.5 wt% MWCNTs, therefore, provides a zT of ~ 1.5 at 573 K, and an average zT of 1.14 in 323 to 573 K. Moreover, this composite has provided the energy conversion efficiency of approximately 8.1% in a single leg thermoelectric generator at temperature difference of 283 K. This is an ideal candidate of low-grade heat recovery due to such findings.

Besides the actual performance of the material, the work has also uncovered a compositing scheme with which to construct future generations of thermoelectric materials, by exploiting huge phonon-spectrum mismatch between the matrix and additive phases. Besides the reinforcement of zT at room temperature, such an adoption provides a next step towards the rational synthesis of high-performance thermoelectric composites. The recent researchers approve the optimization of the thermo-behavior of the CNTs through the regulation of the character of the contacts between the CNTs and the additional materials as the large part of the thermal attributes of the CNTs is their response to increasing and diminishing thermal loadings [47].

It is demonstrated that carbon nanotubes (CNTs) may sustain both the structure and the heat performance under dynamic thermal cycling, which makes them especially appealing to be used in the areas with rapidly varying temperatures. The aim of this experiment is to come up with a personal cooling device using PCM that is safe, comfortable and convenient to use every day. Wearable coulings such as PCMs embedded vests can also effectively eliminate wearer discomfort and thermal stress caused by heat. Regarding the design proposed, the thermal storage materials are chosen as Polyethylene Glycol-600 (PEG-600) and thermally accelerated Glauber salt, with the melting temperature of 18.2 °C and 27.0 °C, respectively. Under Differential Scanning Calorimetry (DSC), the thermal stability of the GO/PEG-600 mixture and the extrapolated Glauber salt were studied and as a result, the values of the latent heats and phase transition temperature range could be found. These results confirm the feasibility of the obtained materials to be incorporated into personal cooling garments that are aimed at regulating the temperature in the daily setting. Adding 0.6 wt.% of graphene oxide to PEG-600, the thermal conductivity improves by 6.1 percent and the latent heat enhances by 8 percent. The phase change materials are placed in aluminum containers, closed and stored in a refrigeration unit. The cooling vest has already been developed and can take up to 95 min during cooling. The cooling vest is recharged by placing the vest in a temperature that is below 10 °C, within 20 min. The proposed personal cooling system is a passive, low cost, and safe cooling system, which cools the human body [48]. Their effective implementation into actual life applications depend on the knowledge of their inherent characteristics and also the knowledge about how they behave when in contact with other constituents in composite systems (Fig. 2.11).

Other features of CNTs that can be desirable to high performance applications besides their great thermal conductivities are good thermal stabilities. They owe their stability to high temperature to the sp2 hybridization of the structure that renders them stable even with high temperature without significant degradation [49]. It means that even CNTs can be employed in high temperature applications, e.g. in a combustion system or the aerospace sector, where there is no performance compromising drop. The details of the outlined uses of CNTs are in the option of conducting heat and the ability to move electrons relatively freely and, therefore, are of much interest in coming up with high-tech

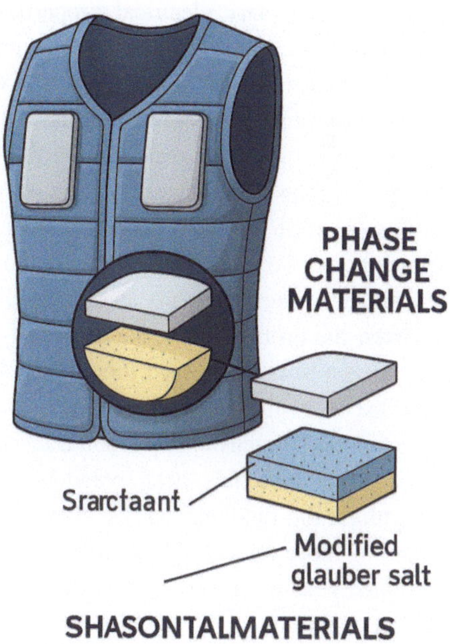

Fig. 2.11 Development of a wearable thermal regulation system using GO/PEG-600 Nanocomposite and thermally enhanced Glauber salt-based PCMs

electric gadgets and sensors. Scientists are also more accurately able to modify their thermal conductivities in innovative varieties of carbon nanotubes (CNTs) functionalization, hence become more applicable in specific sectors, including photothermal treatment and energy storage systems [50]. This provides carbon nanotubes with a lot of prospects in nanotechnology and materials science, namely, in thermal applications.

2.4 Concluding Remarks

Carbon nanotubes (CNTs) are highly flexible and promising type of nanomaterials, which possess remarkable mechano-physical stability, electron conduction capability, and heat transfer capability. The intricate combination of their structural parameters, i.e., chirality, aspect ratio, wall count and surface functionalization, and their consequent physical property combinations indicates the necessity of additional classification and engineering efforts. Acting as multifunctional reinforcers in composite materials, CNTs reinforces greatly the mechanical strength, electrical conductivity, thermal stability, and the overall durability. Such developments are vital to a number of more complex uses such as aeronautical structures, nanoelectronics, medical equipment, and energy storage. Such a property as the ability to tailor CNT properties through dispersion control and surface modification improves their utilization, allowing to develop high-tech multifunctional materials.

The extraordinary electrical properties of carbon nanotubes owing to their tunable electronic band structures and dimensions have made them to be useful components in nanoelectronics devices, biosensors, and optoelectronic systems. On-going improvement in synthesis processes, as well as the combination of CNTs with other active materials will be the key towards realizing the full potential of these materials in future electrical designs. Carbon nanotubes (CNTs) have excellent strength, stiffness, and flexibility that surpass those of the conventional materials and enable new high-performance structural composites to be developed. It has demonstrated great enhancements in the mechanical performance when they are incorporated into polymer, ceramic, and metallic matrices; however, the problem of scalability, cost-effective production, and environment friendliness remain to be addressed thoroughly. In terms of thermal property, carbon nanotubes show excellent thermal conductivity owing to well-ordered sp 2 -bonded carbon lattice and quasi one-dimensional phonon transportation mechanism. The roles they play in thermal interface materials, heat spreaders and high technology composite systems all serve to highlight the fundamental role they play in thermal management which is increasingly becoming more critical in high density electronics and energy devices. It is important that continued research work be done to develop our understanding of how CNT behave in various operating conditions, how their interactions can be optimized in composite matrices and how the various barriers to their large-scale industrial application can be broken. Improvements in accuracy of synthesis, methods of functionalization and compatibility with existing manufacturing processes will be key towards harnessing the entire potential of CNTs.

Carbon nanotubes (CNTs) are a part of an exceptional scope of electrical, mechanical, and thermal properties which is subjected to one dimensionally confined and quantum-mechanical form, and lattice established on graphene.

- The carbon nanotubes (CNTs) may be considered as electrically behaving in either of the following manners–metallic-type nano-wires or semiconducting channels with adjustable bandgaps.
- The CNTs are stronger in strength to weight and elastic resilience as compared to conventional materials.
- Their excellent thermal conductivity provides an opportunity to use superior technologies of heat management.
- In ongoing research, the goal consists of trying to:
- Ability to enhance regulation of synthesis (synthesis regulation: chirality, length, defect density)
- Establish courses of scaleable manufacturing in device integration
- Utilize the properties of the CNT in the other emerging areas like:
- The stretchy electronics
- Energy gathering equipment's
- Composites of the next generation.

An understanding of, and utilization of the correlation between structure and properties will be key to developing CNT based technology. To sum up, carbon nanotubes are the most important objects of contemporary materials science and nanotechnology. Their unique and adaptive nature makes them enablers of huge technical advancement in many fields. With the growth of multidisciplinary collaborations, future holds considerable prospects to CNT based developments, which will incorporate ground-breaking utilizations, which will affect next-generation materials science, nanoelectronics, and energy technology.

References

1. Nath, K., Ghosh, S., Katheria, A., Das, P., Chowdhury, S., Hazra, P., Azam, S., & Das, N. (2022). Design of interconnected 1d nanomaterials with selective localization in biodegradable binary thermoplastic nanocomposite films for effective electromagnetic interference shielding effectiveness. Polymers for Advanced Technologies, 34(3), 1019–1034. https://doi.org/10.1002/pat.5949
2. Ahmadi, H., Khalaj, G., Najafi, A., Abbasi, S., & Safari, M. (2022). Metakaolin-red mud/carbon nanotubes geopolymer nanocomposite: mechanical properties and structural studies. Materials Research Express, 9(2), 025011. https://doi.org/10.1088/2053-1591/ac54d6

3. Moresco, S., Siciński, M., Gozdek, T., Giovanela, M., Carli, L., Crespo, J., & Bieliński, D. (2024). Plasma treatment of rice husk ash for application in tire retread compounds based on natural rubber. Polymer Engineering & Science, 65(2), 700–713. https://doi.org/10.1002/pen.27035
4. RamReddy, C. and Saran, H. (2022). Stability analysis of inclined mhd and joule effects in ti-alloy and mwcnt/water hybrid nanofluid: flow separation. Proceedings of the Institution of Mechanical Engineers Part C Journal of Mechanical Engineering Science, 237(6), 1486–1498. https://doi.org/10.1177/09544062221126647
5. Tanmathusorachai, W., Aulia, S., Rinawati, M., Chang, L., Chang, C., Huang, W., Lin, M.-H., Su, W.-N., Yuliarto, B., & Yeh, M. (2024). High-entropy prussian blue analogue derived heterostructure nanoparticles as bifunctional oxygen conversion electrocatalysts for the rechargeable zinc–air battery. ACS Applied Materials & Interfaces, 16(45), 62022–62032. https://doi.org/10.1021/acsami.4c13387
6. López-Moreno, A., Villalva, J., & Pérez, E. (2022). Mechanically interlocked derivatives of carbon nanotubes: synthesis and potential applications. Chemical Society Reviews, 51(23), 9433-9444. https://doi.org/10.1039/d2cs00510g
7. Mahaling, B., Low, S., Ch, S., Addi, U., Ahmad, B., Connor, T., Mohan, R.R., Biswas, S. & Chaurasia, S. (2023). Next-generation nanomedicine approaches for the management of retinal diseases. Pharmaceutics, 15(7), 2005. https://doi.org/10.3390/pharmaceutics15072005
8. Xiong, J. and Ma, N. (2022). Transcriptomic and metabolomic analyses reveal that fullerol improves drought tolerance in brassica napus l. International Journal of Molecular Sciences, 23(23), 15304. https://doi.org/10.3390/ijms232315304
9. Bu, Y. and Liu, Y. (2024). Investigating the impact of geometric dimensions on the sensitivity of porous flexible capacitive pressure sensors., 54. https://doi.org/10.1117/12.3050268
10. Agarwala, A., Tahsin, T., Ali, M., Sarker, S., Abhi, S., Das, S., Hasa, M., Tasneem, Z., Islam, M., Islam, R., Badal, F.R., & Ahamed, M. (2024). Towards next generation power grid transformer for renewables: technology review. Engineering Reports, 6(4). https://doi.org/10.1002/eng2.12848
11. Jinkins, K., Dwyer, J., Suresh, A., Foradori, S., Gopalan, P., & Arnold, M. (2023). Para-meters affecting interfacial assembly and alignment of nanotubes. Langmuir, 39(40), 14433–14440. https://doi.org/10.1021/acs.langmuir.3c02000
12. Tousignant, M., Ourabi, M., Niskanen, J., Mirka, B., Bodnaryk, W., Adronov, A., & Lessard, B. (2022). Poly (ionic liquid) dielectric for high performing p- and n-type single walled carbon nanotube transistors. Flexible and Printed Electronics, 7(3), 034004. https://doi.org/10.1088/2058-8585/ac928f
13. Metternich, J., Hill, B., Wartmann, J., Ma, C., Kruskop, R., Neutsch, K., Herbertz, S., & Kruss, S. (2024). Signal amplification and near-infrared translation of enzymatic reactions by nanosensors. Angewandte Chemie, 63(9). https://doi.org/10.1002/anie.202316965
14. Zheng, D., Yi, W., Zhou, J., Hou, J., Si, J., & Hou, X. (2023). Two-component polymer sorting to obtain high-purity s-swcnts for all-carbon photodetectors. Chemistry—An Asian Journal, 18(21). https://doi.org/10.1002/asia.202300651
15. Figarova, S. R., Aliyev, E. M., Abaszade, R. G., Alekberov, R. I., & Figarov, V. R. (2021). Negative differential resistance of graphene oxide|sulphur compound. *Journal of Nano Research*, 67, 25–31. https://doi.org/10.4028/www.scientific.net/JNanoR.67.25
16. Nakanishi, Y., Furusawa, S., Sato, Y., Tanaka, T., Yomogida, Y., Yanagi, K., Zhang, W., Nakajo, H., Aoki, S., Kato, T., Suenaga, K., & Miyata, Y. (2023). Structural diversity of single-walled transition metal dichalcogenide nanotubes grown via template reaction. Advanced Materials, 35(46). https://doi.org/10.1002/adma.202306631

17. Shrestha, S., Parajuli, S., Park, J., Yang, H., Cho, T., Eom, J., Cho, S.-K., Lim, J., Cho, G., & Jung, Y. (2023). Improving stability of roll-to-roll (r2r) gravure-printed carbon nanotube-based thin film transistors via r2r plasma-enhanced chemical vapor-deposited silicon nitride. Nanomaterials, 13(3), 559. https://doi.org/10.3390/nano13030559
18. Abaszade, R. G., Mammadov, A. G., Bayramov, I. Y., Khanmammadova, E. A., Kotsyubynsky, V. O., Gur, E. Y., Kapush, O. A., & Boychuk, V. M. (2022). Structural and electrical properties of sulfur-doped graphene oxide/graphite oxide composite. *Physics and Chemistry of Solid State*, 25(2), 256–260. https://doi.org/10.15330/pcss.23.2.256-260
19. Mammadov, A. G., Abaszade, R. G., Kotsyubynsky, V. O., Gur, E. Y., Bayramov, I. Y., Khanmammadova, E. A., & Kapush, O. A. (2022). Photoconductivity of carbon nanotubes. *Technical and Physical Problems of Engineering*, 14(3), 155–160.
20. Abaszade, R. G., Mammadov, A. G., Khanmamedova, E. A., Aliyev, F. G., Kotsyubynsky, V. O., Gür, E., Soltabayev, B. D., Margitych, T. O., Stetsenko, M. O., Singh, A., & Arya, S. (2024). Photoconductivity of functionalized carbon nanotubes. *Digest Journal of Nanomaterials and Biostructures*, 19(2), 837–843. https://doi.org/10.15251/DJNB.2024.192.837
21. Abaszade, R. G., Mammadov, A. G., Bayramov, I. Y., Khanmammadova, E. A., Kotsyubynsky, V. O., Gur, E. Y., & Kapush, O. A. (2022). Modeling of Voltage-Ampere characteristic structures on the basis of graphene oxide/sulfur compounds. *Technical and Physical Problems of Engineering*, 14(2), 302–306.
22. Metternich, J., Sistemich, L., Nißler, R., Herbertz, S., & Kruss, S. (2023). Rational design of near-infrared fluorescent carbon nanotube biosensors with covalent DNA-anchors. https://doi.org/10.26434/chemrxiv-2023-838d6
23. Giesler, J., Weirauch, L., Rother, A., Thöming, J., Pesch, G., & Baune, M. (2023). Sorting lithium-ion battery electrode materials using dielectrophoresis. ACS Omega, 8(29), 26635–26643. https://doi.org/10.1021/acsomega.3c04057
24. Selamat, M., Zainal, M., Yasin, M., Hamzah, N., & Sidik, N. (2023). Modelling of the flame synthesis of single-walled carbon nanotubes in non-premixed flames with aerosol catalyst. ARNHT, 13(1), 39–51. https://doi.org/10.37934/arnht.13.1.3951
25. Saeed, A., Abbas, M., El-Hawary, W., Issa, Y., & Singh, B. (2022). A core–shell au@tio2 and multi-walled carbon nanotube-based sensor for the electroanalytical determination of H_2O_2 in human blood serum and saliva. Biosensors, 12(10), 778. https://doi.org/10.3390/bios12100778
26. Okotrub, A., Gorodetskiy, D., Гусельников, A., Kondranova, A., Bulusheva, L., Korabovska, M., Meija, R., & Erts, D. (2022). Distribution of iron nanoparticles in arrays of vertically aligned carbon nanotubes grown by chemical vapor deposition. Materials, 15(19), 6639. https://doi.org/10.3390/ma15196639
27. Guvalov, A. and Mammadov, A. (2024). Preparation of effective nanomodified polymer-bitumen adhesives for asphalt concrete. E3s Web of Conferences, 515, 01014. https://doi.org/10.1051/e3sconf/202451501014
28. Heriyanto, A., Okamoto, N., Cho, Y., Abe, R., Okamoto, C., Pandey, M., Benten, H., & Nakamura, M. (2025). Influence of polyoxyethylene stearyl ether and subsequent annealing on n-type doping of cnt yarns for flexible thermoelectric generators. ACS Omega. https://doi.org/10.1021/acsomega.4c10388
29. Krishnan, H., Gopinath, S., Adam, T., Fakhri, M., Salim, E., & Patil, N. (2024). Molecular-imprinting assisted polydopamine-aptasensor on carbon and gold nanomaterials construct for the haemophilia b biomarker detection. IJNeaM, 17(June), 25–30. https://doi.org/10.58915/ijneam.v17ijune.831
30. Radivoievych, A., Kolp, B., Grebinyk, S., Prylutska, S., Ritter, U., Zolk, O., Glokler, J., Frohme, M., & Grebinyk, A. (2022). Silent death by sound: c60 fullerene sonodynamic treatment of cancer cells. https://doi.org/10.21203/rs.3.rs-1193610/v1

31. Serrato-Mireles, I., Agarwal, V., Arredondo-Vargas, E., Garnica-Romo, M., López-Gómez, R., & Schubert, N. (2025). Phytoremediation of heavy element contaminated water using eichhornia crassipes (water hyacinth) injected with multiwalled carbon nanotubes and titanium dioxide nanotubes—a clean environmental nanotechnology. https://doi.org/10.21203/rs.3.rs-6331778/v1
32. Azizi, S., Askari, M., Moghadam, M., Seifi, M., & Bartolomeo, A. (2023). Ni3s4/nis/rgo as a promising electrocatalyst for methanol and ethanol electro-oxidation. Nano Futures, 7(1), 015002. https://doi.org/10.1088/2399-1984/acb02b
33. Lajoie, J., Barkley, S., & Sippel, T. (2023). High-speed, microwave ignition: antenna igniters with frequency and bandwidth selectivity. Propellants Explosives Pyrotechnics, 48(8). https://doi.org/10.1002/prep.202200347
34. Murjani, B., Kadu, P., Bansod, M., Vaidya, S., & Yadav, M. (2022). Carbon nanotubes in biomedical applications: current status, promises, and challenges. Carbon Letters, 32(5), 1207–1226. https://doi.org/10.1007/s42823-022-00364-4
35. Salehcheh, M., Zeidooni, L., Dehghani, M., Alboghobeish, S., & Shirani, M. (2023). Effects of caffeic acid on multi-walled carbon nanotubes -induced mitochondrial toxicity in rat kidney. Jundishapur Journal of Natural Pharmaceutical Products, 18(2). https://doi.org/10.5812/jjnpp-133171
36. Ibrahim, N., Sameh, H., & Amer, I. (2023). The impact of nanoparticles implementation on conservation of heritage buildings (overview the nanocathedral project). International Journal of Advanced Scientific Research and Innovation, 6(1), 38–89. https://doi.org/10.21608/ijasri.2023.207409.1004
37. Chen, J. (2022). Effects of different factors on the heat conduction properties of carbon films and fibers. East European Journal of Physics, (2), 91–98. https://doi.org/10.26565/2312-4334-2022-2-11
38. Chen, J. (2022). Effects of structure, purity and alignment on the heat conduction properties of carbon nanotubes and fibers. Muş Alparslan Üniversitesi Fen Bilimleri Dergisi, 10(1), 909–915. https://doi.org/10.18586/msufbd.1065544
39. Alshihabi, M. and Kayacan, M. (2024). Effect of nanosized carbon nanotubes, titanium nitride and cubic boron nitride powders on mechanical and thermal properties of sla 3d printed resin composites. Polymer Composites, 45(17), 15561–15573. https://doi.org/10.1002/pc.28850
40. Abaszade, R. G., Babanli, M. B., Kotsyubynsky, V. O., Mammadov, A. G., Gür, E., Kapush, O. A., Stetsenko, M. O., & Zapukhlyak, R. I. (2023). Influence of gadolinium doping on structural properties of carbon nanotube. *Physics and Chemistry of Solid State, 24*(1), 153–158. https://doi.org/10.15330/pcss.24.1.153-158
41. Figarova, S. R., Aliyev, E. M., Abaszade, R. G., & Figarov, V. R. (2023). Negative thermal expansion of sulphur-doped graphene oxide. *Advanced Materials Research*, 1175, 55–62. https://doi.org/10.4028/p-rppn12
42. Abaszade, R. G., Aliyev, E. M., Mammadov, A. G., Khanmammadova, E. A., Guliyev, A. A., Aliyev, F. G., Zapukhlyak, R. I., Budak, H. F., Kasapoglu, A. E., Margitych, T. O., Singh, A., Arya, S., Gür, E., & Stetsenko, M. O. (2024). Investigation of thermal properties of gadolinium doped carbon nanotubes. *Physics and Chemistry of Solid State*, 25(1), 142–147. https://doi.org/10.15330/pcss.25.1.142-147
43. Mammadov, A. G., Abaszade, R. G., Babanli, M. B., Kotsyubynsky, V. O., Gur, E., Soltabayev, B. D., Margitych, T. O., & Stetsenko, M. O. (2023). Photoconductivity of gadolinium-doped carbon nanotubes. *International Journal on Technical and Physical Problems of Engineering*, 15(3), 53–58.
44. Ivanichok, N., Kolkovskyi, P., Ivanichok, O., Kotsyubynsky, V., Boychuk, V., Rachiy, B., Bembenek, M., Warguła, Ł., Abaszade, R., & Ropyak, L. (2024). Effect of thermal activation on

the structure and electrochemical properties of carbon material obtained from walnut shells. *Materials*, 17(2514), 1–21. https://doi.org/10.3390/ma17112514

45. Song, W., Sun, C., Yang, K., Ke, S., Li, X., Liu, M., Sun, C., Zhao, W., & Zhang, Q. (2024). Magnetism-induced diffuse scattering effect triggers excellent thermoelectric performance. Energy & Environmental Science, 17(14), 5216–5226. https://doi.org/10.1039/d4ee01954g
46. Qin, Z., Li, D., Ou, Y., Du, S., Jiao, Q., Peng, J., & Liu, P. (2023). Recent advances in polydopamine for surface modification and enhancement of energetic materials: a mini-review. Crystals, 13(6), 976. https://doi.org/10.3390/cryst13060976
47. Liang, Z., Xu, C., Shang, H., Ning, M., Tong, T., Song, S., Ren, W., Shi, Z., Liu, X., Ding, F., Bao, J., Wang, D., & Ren, Z. (2023). Near-room-temperature thermoelectric performance enhancement via phonon spectra mismatch in $Mg_3(Sb, Bi)_2$-based material by incorporating multi-walled carbon nanotubes. Advanced Energy Materials, 13(25). https://doi.org/10.1002/aenm.202301107
48. Farhan, F. and Hamza, M. (2024). Use of modified organic and inorganic phase change materials for the development of personal cooling device. Materials Science Forum, 1129, 29–36. https://doi.org/10.4028/p-v06hrd
49. Mohammadnejad, S., Mohammadi, M., Ghartavol, H., & Askari, M. (2023). Lithium-functionalized graphene oxide with a low load of pt as an efficient counter electrode for dye-sensitized solar cells. Journal of Applied Electrochemistry, 54(2), 275–287. https://doi.org/10.1007/s10800-023-01967-z
50. Niu, Q., Sun, Q., Bai, R., Zhang, Y., Zhuang, Z., Zhang, X., Xin, T., Chen, S., & Han, B. (2022). Progress of nanomaterials-based photothermal therapy for oral squamous cell carcinoma. International Journal of Molecular Sciences, 23(18), 10428. https://doi.org/10.3390/ijms231810428

3 Application Carbon Nanotube-Based Electronics

Carbon nanotubes (CNTs) have turned out to be the important components in the nanotechnology, especially due to their unique electrical, thermal, and mechanical properties. These cylindrical nanostructures were discovered by Iijima in 1991 and since then intensive research has been motivated by taking advantage of their properties to apply in various fields, most notably electronics and materials science. Bioimaging in the near-infrared-II (NIR-II) wavelength is emerging as a mandatory visualization modality in real-time exploration in fundamental biology studies and therapies [1].

In addition, the versatility and relevance of CNTs are further demonstrated by the constant progress in the field of their medicinal use. Drug delivery systems are also being explored using carbon nanotubes because of the ability to permeate through cell membranes coupled with the ability to transport therapeutic agents successfully and therefore may lead to better efficacy of treating diseases such as cancer. The interdisciplinary activity concerning the use of CNTs also continues to grow, combining fields like nanomedicine and biotechnology [2–5].

The increased importance of carbon nanotubes in electronics, environmental science, and health contribute to the strengthening of individual disciplines and the collaboration of disciplines. Such inherent properties of carbon nanotubes as electrical conductivity, thermal resistance, and chemical stability have enabled their engagement in various technological innovations in multiple industries [6, 7]. This advancement indicates that further research is necessary to explore the entire possibility of the carbon nanotube, especially in the interest of the environment and human well-being.

To conclude, carbon nanotubes can be considered as the crown of the nanotechnology development representing a combination of the hope and utilization in a number of applications. Their unusual characteristics give ample opportunities to enhance electronic

R. G. O. Abaszade and A. K. Haghi, *Carbon Nanotube Devices for Nanoelectronics*,
Synthesis Lectures on Solid State Materials and Devices,
https://doi.org/10.1007/978-3-032-02114-4_3

technologies and solve urgent global problems in medicine and environmental protection [8, 9]. The ongoing exploration of carbon nanotubes in numerous domains is likely to give important findings, to promise a future where the enormous potential of these nanostructures will be exploited to benefit society.

3.1 Introduction

One of the largest properties of nano tube of carbon is electrical conductivity and thus can be utilized in nanoelectronics. The metallic or semiconducting property of carbon nanotubes, which is based on their chirality and diameter, creates the path to the widespread application of CNTs in transistor, diodes, and sensors [10, 11]. They are also high aspect ratio and thermally stable, further used in a large number of other areas, such as high-performance computing, flexible electronics, as well as energy conversion systems [12, 13]. It has discussed the improvement of the properties of carbon nanotubes with Graphene in several ways of doping and composite materials. To a significant extent, electrical and optical characteristics of CNTs could be modified upon introduction of such elements as nitrogen and metals and such product can be more practically used in a range of applications [14]. Through such adaptation, the optoelectronics development can be provided where carbon nanotubes can offer to be the light source or the photodetector and consequently the target of the interest of the scientists who attempt to develop the solar energy technology and environmental sensors [15].

The aspect of sustainability regarding the use of carbon nanotubes is colossal. It was further found that the carbon nanotubes (CNTs) is likely to be effective in environmental remediation particularly in water purification technologies due to its high capacity of adsorption and large surface area that may be adopted in many aspects to remove contaminants and poisons beside water purification technologies [16]. The application is one of the potential solutions in the water quality issue and it is considered an example of the carbon nanotubes contribution to the sustainable activities promotion through advanced materials, i.e., the chance of clean industries [17]. Carbon nanotubes are also peculiar since they make use of two of the largest assets of materials science, and they cannot be overlooked on the road leading to the achievement of a sustainable technology. This has made it highly essential to use when employed in the application on electronic devices where efficiency and reduced cost of operating the device are of real concern. The geometry of the field emitter: the carbon nanotubes (CNTs); the emission characteristics are strongly anisotropic to the CNTs structure: high currents could be emitted by closely packed arrays; however, a screening effect could also occur and the emission efficiency would decrease [18].

As well as the structure, CNTs chemical modifications and doping mean have been explored profoundly in the quest to enhance CNTs emission properties. Doping of carbon nanotubes via metal could be used to improve the field emission since the process affects the work functions and augments the density of charge carriers with the use of compounds like nickel [19]. In addition, it is also desirable to have a narrow control range over the electric field, to avoid electrical arcing at least a side effect of which may be severe degradation of the CNT emitters. Even the fields emission may in fact be a factor of the geometry of CNT arrays. The upstanding direction of CNT electron emission has been depicted to possess higher electron emission because of attractive shape that strengthens the electric field in the tip [20]. Synergetic effects of other hybrid schemas, in particular, CNTs combined with a large number of materials, including diamond, have shown a more stable and efficient field emission, which reflects potential areas of employment device architecture [21]. The development of such advanced designs has a future in next generation applications (particularly in medical imaging where during normal operating conditions its high production of electrons are required to remain stable).

Moreover, the fact that new processes on how carbon nanotubes (CNTs) should be synthesized primarily the adoption of chemical vapour deposition (CVD) processes will be facilitated to manufacture quality CNTs to accommodate a specific field emission. The structural quality is governed by maximizing the growth conditions and it has to be present prior to attaining a suitable electrical quality to increase the emissions performance [22]. It is to be hoped that a supported emphasis on synthesis, growth processes, can only increase the ability of the field to utilize CNTs in advanced validation of electronic applications.

3.2 Application of CNTs in Field Emission

Carbon nanotubes (CNTs) are of distinctive structural features and excellent electrical properties that makes them focal points in the field emission technology hence influencing various applications of the same such as electron sources and sophisticated imaging systems among others. CNTs have multifunctional characteristics attributed to high aspect ratio, high electrical conductivity and the ability to permit the establishment of field electron emission at low potentials hence endowing them with important benefits over conventional materials. Current studies have explained that alignment and change in the surface is a major structural feature in promoting their radiation strength. The stability and performance of the enhancing substrate can just as well be boosted, provided that the methods enable to achieve unified at least approximate efficiency or the supporting strain [23–25]. On the backdrop of such discoveries, there are implication in the fabrication of structured substrates wherein efficient electron emission can be obtained and at the same time retaining a small form factor which can be used in portable electronics.

Field emission refers to the emission of electrons under the influence of a strong electric field and these may be emitted by a substance surface (Table 3.1). Owing to:

- High aspect ratio Implies higher localized electric fields
- Wonderful conductivity of electricity;

Table 3.1 Advantages of CNTs in field emission

Property	Benefit for field emission
High aspect ratio	Strong local field enhancement
Low work function (~4.5 eV)	Low turn-on voltage
High thermal conductivity	Stable operation under high current
Chemical inertness	Long lifetime of emitters

- Chemical and mechanical hardiness.

Field emission current density (Fowler–Nordheim equation):

$$J = \frac{A}{\varphi} E^2 exp\left(-\frac{B\varphi^{\frac{3}{2}}}{E}\right)$$

where:

- J—current density;
- E—applied electric field;
- φ—work function of the material;
- A, B—constants.

CNT arrays exhibit:

- Low turn-on fields (~1–5 V/μm);
- High current stability.

Application in:

- Flat-panel displays;
- Electron sources for X-ray tubes;
- Microwave amplifiers.

To sum up, the ongoing research on lots of applications of carbon nanotubes in field emission proves one can extremely count on this material to lead a lot of changes to many electronic devices. Synergistic combination of their structural, electrical and thermal properties does hold promise of innovation that can lead to major advancements in the field of electron sources as well as other areas. Current research shows that there is a complex relationship between the properties of carbon nanotubes (CNTs) and their efficacy in emission scenarios and this explains why the future looks bright of the use of CNTs in high-end electronics and other state-of-the-art applications.

3.3 Application of CNTs in Data Storage

Carbon Nanotubes (CNTs) has appeared as a transformative material in many fields, most notable as energy storage material. This discussion will focus on the numerous applications of carbon nanotubes to energy storage technologies by compiling findings of different academic literature and discussing the implications that it has towards energy systems in the future. Both of the two variations have excellent mechanical properties, where tensile strengths are up to 130 GPa, and the moduli are larger than 1 TPa [26, 27]. It is important that thermal conductivity of fluids in various heat exchangers be regulated. In this regard, carbon nanotubes (CNTs) can enhance the heat transfer abilities using the fact that they have excellent thermal conductivity. The performance of a modified nanofluid comprised of water and SWCNT and MWCNT is observed depending on the proportion of CNTs. The addition of carbon nanotubes (CNTs) to a fluid lowers the temperature and velocity of a fluid and all these properties make them useful in a cooling system. Figure 3.1 shows that the highest heat transfer with plates is 35.35 percent and 25.35 percent where 0.01 and 0.02 are 0.1 and 0.2 respectively. The fluid inertial forces and inertial pressure regime substantially decrease the velocity and the thermal buoyancy forces promote increase in flow. Mechanical characteristics of CNTs allow their incorporation into composite materials, which significantly enhance both thermal and mechanical characteristics of such composites, hence enhancing the overall performance of the equipment by the composites [28].

The main advantage of the use of CNTs in the energy storage system is that they possess greater electrical conductivity and thermal conductivity. The characteristic facilitates formation of conductive networks within electrode materials strengthening energy transfer and ion diffusion in the process of charging and discharging. It has been studied

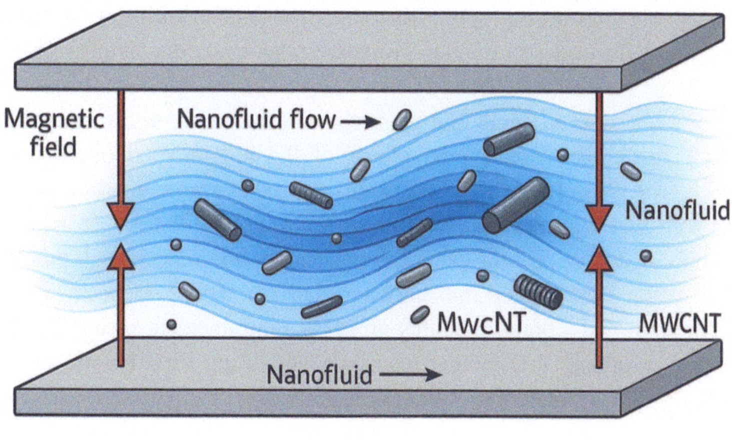

Fig. 3.1 Diagram of oscillatory hybrid nanofluid flow

that carbon nanotubes (CNTs) provide a highly conductive framework in the scheme of sodium-ion batteries, which facilitates the improved charge transfer and promotes Na-ion diffusion [23]. The speed and effectiveness of energy storage activities are dictated by the rapid transit of ion. Supercritical overuse of nanotubes (CNTs) helps to significantly increase specific capacitance and cycling stability. MXene/N-CNT composites are a good example of using transistors in the fabrication of supercapacitors with higher energy densities than on the traditional materials, evident of the capacity that CNTs have in addressing the expanding energy storage needs that technology demands [29, 30]. It is due to the impressive surface area and low internal resistance that contributes to the effectiveness of supercapacitors [31] carbon nanotubes would add energy storage capabilities to the efficiency of supercapacitors. Moreover, CNTs can be used as electrocatalysts of hybrid systems. Research proves that the synergy of carbon nanotubes with transition metals, including copper, maximizes and promotes their catalytic ability, making them more useful in improving the electrochemical energy storage system [32, 33]. This partnership can assist in developing strong batteries and fuel cells, which are essential in the transition of renewable energy systems.

CNTs are being studied as the method of data storing due to their:

- Nano-scale storage space is very dense.
- Conductivity is high hence fast transfer of charge.
- Longevity in mechanical storage device is offered by mechanical strength.

Applications:

1. NVM (non-volatile memory) based on CNTs - > CNT-FET memories are constructed using CNT-based NVM as channel material
2. resistive RAM (ReRAM) using CNTs switch (controlled by localized
3. Magnetic storage has been proposed as a direction to take, with carbon nanotubes used as spacers and electrodes in magneto resistive random-access memory (MRAM).

The hybrid energy storage, including aluminum-ion batteries, is also greatly facilitated with CNTs because they modify the places of the ion interactions hence increasing the storage power [34]. Those hybrid systems can offer viable solutions to the current lithium-ion tech that will ease the limit of the conventional battery. In addition to that, CNTs structural materials present themselves as energy boosters through their mechanical enhancement when incorporated into polymer matrixes. This evidenced improvement in the tensile strength and storage modulus of CNTs reinforcement means that composites are capable of sustaining mechanical loads without losing their ability to store energy successfully [35, 36]. Connection of carbon nanotubes with smart and self-sensing materials is an emerging approach toward effective energy consumption and efficiency in the evolving energy storage environment (Table 3.2).

Table 3.2 Overview of Ni–carbonaceous composite coatings and research status

Nickel-based composite coating	Carbonaceous reinforcement	Enhanced properties	Current research focus	Missing characterizations	Application areas
Ni/diamond	Diamond nanoparticles	Hardness ↑, Wear resistance↑, Corrosion resistance ↑	Microstructure, Phase formation	Electrochemical, magnetism, tribology	Aerospace, defense
Ni/CNT	Carbon nanotubes	Toughness ↑, lubrication ↑, grain refinement	Deposition parameters, Microstructure	Magnetism, advanced tribology	Automotive, electronics
Ni/graphene	Graphene nanosheets	Corrosion resistance ↑, Mechanical stability ↑	Surface morphology, Electrical behavior	Magnetism, tribology	Aircraft, coatings
Ni/C_{60}	C60 (Fullerene)	Potential for biocompatibility, catalysis	Not yet explored	All fields: structure, corrosion, magnetism	Biomedical, Catalysis
Ni/carbon black	Amorphous carbon	Lubrication, cost-effective	Some basic deposition studies	Surface bonding mechanisms	low-cost industries

Although CNTs have several advantages, challenges surrounding their production, scaling and combination with other materials are still there. To increase the competency of carbon nanotubes in a variety of applications, innovations in the synthesis methods, e.g., started chemical vapour deposition, are under analysis [37].

CNTs have numerous applications in energy storage that marks a significant stake in materials science and engineering. The application of carbon nanotubes has changed the paradigm of energy preservation, hence changing paradigm of traditional battery performance and creation of new energy solutions. Conclusively, CNTs in energy storage can help depict how new materials might radically change technology through their exceptional properties. Continued funding towards the research and development of CNT-based technologies would enhance their adoption in the various platforms that address the global challenges of energy effectiveness, sustainability as well as storage capacities.

3.4 Application of CNTs in Nanoelectromechanical Systems

The carbon nanotubes (CNTs) are important in the advancement of the nanoelectromechanical systems (NEMS) that have received much interests because of their excellent capability in the mechanical, electrical and thermal aspects. The mechanical properties of the carbon nanotubes are mind blowing and are so light in weight they can prove to be a new application to the nanoelectromechanical systems namely the sensing and actuation devices. Not only does this unique structural framework of CNTs enhance all these attractive properties, but also demonstrates significant benefits in comparison with conventional materials in terms of comparison. The majority of the literature observes that some of the significant contributions in the industry of nanotechnology is made by the allotropes of carbon such as the carbon nanotubes (CNTs) and graphene since they are very simple to incorporate when producing miniature equipment's with the best functionality properties in different contexts which includes sensors and energy gatherers. In a recap, it has been seen that the spectacular nature of carbon nanotubes i.e. high physical strength, conductivity and designable properties makes its application wide in the context of nanoelectromechanical systems. Various approaches of fabrication of MEMS/NEMS using 2D materials have been addressed. The applications in MEMS/NEMS often demand the 2D material arrangements with the specific electric, optical, mechanical properties, whereupon related nanostructures are facilitated by the selective growth and transfer plan. Studio growth, substrate engineering of the single layer crystalline 2D material is one of the significant breakthroughs, as well as a complication of MEMS/NEMS devices. We infer that all of the possible use of such devices can become commercially viable when there is collaboration between academia and industry. Multidomain/transdisciplinary character of MEMS/NEMS that is formed by the interaction of different physical and chemical processes, generates large transformative expenditures of the scientific community and business. Potentiality in terms of creating highly sensitive sensors and actuators with the impregnation of CNTs further proves their potential contribution in the development of the nanotechnology sector. The outstanding member of the enormous family of carbon-based nanostructures, which also includes the physical characteristic of the emerging nanomanufacturing and material research and development activities along the physical characteristics of carbon nanotubes (CNTs), precedes the nanotechnology applications in nanoelectromechanical systems (NEMS). With proper, intact nanoparticle and virus mass spectrometry may prove to be a powerful tool of material science and biology characterization. Single nanoparticle (more than 10MDa) and virus mass can conveniently be measured with nanoelectromechanical system (NEMS) based mass spectrometry without being able to measure with conventional commercially available approaches. In this process charged and isolated molecules of analytes are produced by means of electrospray ionization (ESI) in air and finally introduced in to the NEMS resonator to be ionized and measured. The drawback of NEMS is that they are very small and their surface areas are small hence suboptimal limit-of-detection and capture efficiency. The solution to such limitations to such limitations is by adding an ion lens on the NEMS chip which answers the question. It is argued that under ambient conditions this combined platform enables the detection of gold and polystyrene nanoparticle with two orders-of-magnitude increase in capture efficiency above contemporary state-of-the art. Subsequently, using the technique, we were able to get the mass spectrum of SARS-CoV-2 and BoHV-1 virions. The increase in analytic throughput and the elementary character of the overall design as well as their operability in ambient conditions indicates that NEMS mass spectrometry can be effectively implemented in the direction of the mass spectral detection of designed nanoparticle and biological matter [38, 39]. Carbon nanotubes (CNTs) are the best starting materials of nanoelectromechanical systems (NEMS) due to: Elevated Young's modulus $E \approx 1.0$ TPa.

Low mass High electrical conductivity Representative NEMS devices:

- Nano resonators are oscillators made of carbon nanotube and have high quality factor.
- Nano switches 6 Carbon nanotubes as floating elements in relay networks Sensing: mass, pressure and chemical sensors.

Carbon nanotube based nanoelectromechanical systems: resonant frequency: $f_o = \frac{1}{2L}\sqrt{\frac{E}{\rho}}$.

Where:

- L—length of CNT
- E—Young's modulus
- P—mass density.

The Van der Waals type of bond that makes carbon nanotubes touch the surfaces is also essential in using them in nanoelectromechanical systems. The van der Waals force may

Table 3.3 Properties of CNTs beneficial for NEMS

Property	Benefit for NEMS
High stiffness	High resonant frequencies
Low density	Fast response
High electrical conductivity	Efficient signal transduction
Mechanical resilience	Long operational lifetime

be characterized by Hamaker constant in electromechanical coupling devices in the interaction of carbon nanotubes and electrodes. This model explains how adhesive force affects operating voltage, stability and how these systems respond as a whole mechanically [40].

The main novelty is the accurate detection of temperature value limits at which certain events could be observed on carbon fibers during Joule heating. These readings could allow us to be able to make reading real time calculations of the temperatures of the carbon fibers. Our approach was corroborated by applying the obtained temperature data in synthesizing the CNTs that were surface-localized to one suspended carbon fiber. Raman analysis confirms the increased quality of the structure after the JH, and the resulting hybrid carbon nanotubes/carbon fibers demonstrate the electrical conductivity of up to 476 S/cm, which suggests the potential usage of this method within the development of high-performance carbon products (Fig. 3.3). Moreover, the mechanical characterization of carbon nanotubes, as indicated by the high Youngs modulus, enhance greatly their performance as nanoelectromechanical systems actuators and sensors [41] (Table 3.3).

CNT applications go beyond primitive sensing and they play a pivotal role in activities related to intelligent functionality, such as energy capturing and storing capabilities, which prove to be essential in creating the autonomous NEMS systems of the future. The prerequisite of the functionalities of nanoelectromechanical systems depends on the interaction between the carbon nano tubes and their surrounding surroundings. It is shown in research that the operating efficiency of CNT-based equipment's can be enhanced or hindered considerably through heat and electromagnetic effects. The NEMS thermal dynamics and particularly the topic of Joule heating on CNTs are important to applications in microscopic energy conversion process as well as electronic devices [42].

The design and optimization of CNT-integrated NEMS has been investigated, whereby it is clear that computational analyses and computational modelling frameworks are required to predict device performance in a variety of operating conditions. These materials are quite functional in the aerospace industry, nuclear industry, and the aviation industry since its properties are extraordinary. Relative and continuous modification of its elastic and thermal characteristics over its surfaces de-emphasizes the thermal strain within the material and increases the longevity of the material. One can integrate Eringen assumptions, the Hamiltonian principle and Euler Bernoulli theory to create a memory dependent derivative heat transfer equation with two delay periods of time. Laplace transform was used to obtain the analytical solutions of the domains of the system. To ensure that the suggested model is legitimate, the current model of thermal conductivity was also compared with previous classical and nonclassical thermal conductivity versions [43] (Fig. 3.2 and Table 3.4).

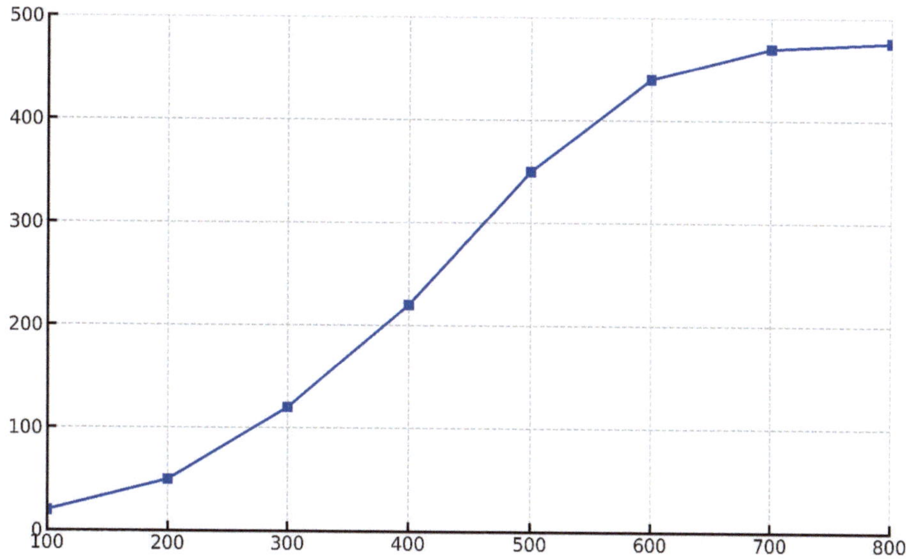

Fig. 3.2 Effect of Joule heating on CNT growth and conductivity of Hybrid CNT/CF structures

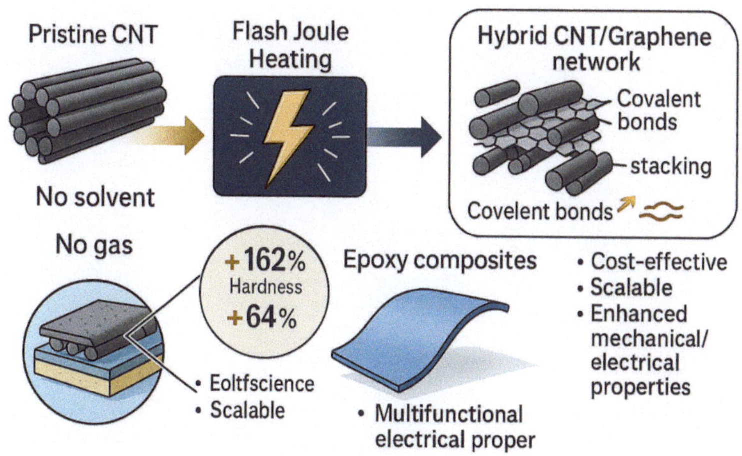

Fig. 3.3 Solvent-free synthesis CNT/graphene hybrids for multifunctional nanocomposites

Table 3.4 Advanced materials and composite coatings for MEMS/NEMS and structural applications

Application domain	Material/composite	Properties enhanced	Current challenges	Opportunities/trends
Structural coatings (aerospace, automotive)	Ni/diamond	Hardness ↑, Wear resistance ↑, corrosion resistance ↑	Electrochemical characterization, magnetism, tribology	Advanced electrochemical analysis, nanoscale tribology
Structural coatings (Automotive, Electronics)	Ni/CNT	Toughness ↑, lubrication ↑, grain refinement ↑	Magnetism, advanced tribology	Multi-scale tribological studies, hybrid integration
Structural coatings (Aerospace, Electronics)	Ni/graphene	Corrosion resistance ↑, Mechanical stability ↑	Magnetism, tribology	Hybrid Ni/Graphene integration, flexible coatings
Functional coatings (Biomedical, Catalysis)	Ni/fullerene	Biocompatibility potential, catalytic activity	Structure, corrosion, magnetism	Novel Ni/Fullerene hybrid coatings
Flexible NEMS/MEMS devices	2D Materials (Graphene, MoS_2, hBN, etc.)	Ultra-thin, flexible membranes, high mobility	Large-scale single-crystal growth, transfer process	Advanced CVD/MBE, dry transfer techniques
NEMS/MEMS devices (CMOS compatible)	Suspended 2D membranes (Graphene, MoS_2)	Low mass, high Q-factors, sensitivity ↑	CMOS integration, packaging	Hybrid CMOS-2D integration, advanced encapsulation
NEMS/MEMS reliability	2D Materials-based MEMS/NEMS	Mechanical stability ↑, Electrical stability ↑	Long-term reliability, scaling	Industry-academic collaboration, standardized testing

In addition, mixing of different nanocomposites like CNTs with other materials, e.g. polymers or metals, may create hybrid products. The properties of hybrid carbon nanomaterial also evince interesting mechanical and electrical functions since such systems are composed of covalent bonding and 2-D layering by means of 2-D stacking of pi–pi interaction. The high cost of the energy required in the production of these kinds of hybrid materials which entail chemical vapour deposition among others, limits the extent of synthesis. Hybrid material proves to show a 162 percent gain in hardness and 64 percent increase in Youngs modulus as compared to unmodified epoxy when triboindenting epoxy composites (Fig. 3.3). Such a hybridization enhances the way the device functions as a whole which allows the creation of multifunctional NEMS according to specific usage [44].

Finally, numerous applications and opportunities of CNTs in nano-electromechanical systems are a rich study area with the potential to develop more. Their mechanical aspects, electrical properties and their abilities to be customized in their functional aspects through structural adaptation make them very significant in the current technological advancement. The possible lines of further research work can be explored in the synthesis of new promising carbon nanotubes and composites, embedding them into complex systems, and commercialization of these technologies to use in numerous fields, as well as in medical diagnosis, environmental surveillance, and high-precision industries.

3.5 Application of CNTs in Super Capacitors

CNTs are becoming critical components in production of supercapacitors due to their impressive electrical conductivity properties, high surface area and mechanical stability. Introduction of CNTs to supercapacitors significantly enhances several performance indicators, such as energy density, as well as capacitance. Their nature has led carbon nanotubes to have increased ion diffusion in the context of charge–discharge cycles and advancements in supercapacitor technology [45, 46]. Another interesting feature of CNTs is their ability to form nanocomposites with a wide variety of materials, and, therefore, enhance the electrochemical activity of supercapacitors. Liu et al. provided an example of this synergy of redox-active chemicals and multiwalled carbon nanotubes (MWCNTs), which resulted in a unique combination of capacitors, which showed considerable capacitance and energy density increases [47]. Functionalization enhances the wettability of CNTs, enhances electrolyte accessibility and interacts more effectively with charge storage through processes of pseudo capacitance [48]. Yang and Zhitomirsky studied the effectiveness of Mn_3O_4-CNTs nanocomposites to find that the conductive nature of CNTs plays a crucial role in the overall capacitance of the asymmetric supercapacitors to dominate the electron transport chain [49].

CNTs with their high surface area and conductivity actually benefit supercapacitors or electrochemical capacitors (Table 3.5). Capacitance equation:

Table 3.5 Performance metrics of CNT-Based supercapacitors

Metric	Typical value with CNT electrodes
Specific capacitance	80–200 F/g
Energy density	5–10 Wh/kg
Power density	10–20 kW/kg
Cycle life	> 100,000 cycles

$$C = \epsilon_r \epsilon_0 \frac{A}{d}$$

where:

- C: capacitance
- ϵ_r—relative permittivity
- ϵ_0—vacuum permittivity
- A—surface area
- D—effective thickness of the double layer

CNTs provide:

- High specific surface area (~1000 m^2/g)
- Excellent electrical pathways
- Stable electrochemical performance.

This trend also includes the further development of hybrid super capacitors, in which conductive polymers such as polyaniline (PANI) synergies with the CNTs to produce significant improvements to charge capacity and charge rates. The combination enhances value of capacitance and ensures structural integrity in the process of cycling [50]. Carbon nanotubes functionalization processes boost their performance in supercapacitors. Plasma treatments can be used to modify the surface through which electrical properties can be benefited as well as increase activation to adsorb ions. The MWCNTs were deposited on stainless-steel mesh in the CVD method where the substrate was directly synthesized and subsequently functionalized by a rapid, solvent-free radiofrequency (R_f) plasma method. The MWCNTs in form of a binder-free, forest-like structure were studied as supercapacitor electrodes, before and during functionalization. The Interstructural and electrochemical characteristics of the electrodes were evaluated. The RF power (1030 W) and the angle of functionalization (120 min) were adjusted so as to achieve maximum capacitance. Functionalization of MWCNT with nitrogen and oxygen significantly increased the capacitive performance and this could be explained by the presence of the functional groups on the MWCNT surface. The highest value capacitance was accomplished in the MWCNT electrode treated using RF plasma under combination of $Ar/C_2H_6/O_2$ gasses at 20 W during 15 min and capturing a specific capacitance of 37.3 1.1 mF cm^2. Studies show that CNTs that are functionalized by using plasma display a better retention of capacitance, and have the potential to stabilize active elements on their surfaces which would therefore allow a higher level of charge storage and stability of the cycle [51]. New developments point out that the efficiency of hybrids supercapacitors similar to those composed of CNTs has increased because of the application of innovational structural organization such as 3D porous structures. These structures make the electrode material have a better ion accessibility and conductivity, which significantly increases the performance of the supercapacitor all-together. Study of materials which have been used to include MXene in CNTs shows that there is the need to develop electrode structures that will enhance the creation of ion diffusion pathways and improve electronic transport efficiency [52]. Also, the conductive CNT additives have substantially improved capacity and reduced overall supercapacitor mass, overcoming limitations of lower capacity and mass of less conductive materials.

Analysis of the mechanical features of CNT structure will reflect other benefits of the use of these materials in flexible and wearable technologies. The scholars have been able to prove enhanced capability to store energy within the wearable electronic devices using imperfective super capacitors composed of CNT additives. This indicates that CNTs are not just good electrochemically but they also have the required mechanical strength with which they can be used in a wide variety of applications due to their interdisciplinary nature [53]. The current approach to the improvement of the electrochemical property of the supercapacitors remains the recruitment of the new processes of which the maximization of the electrode surfaces and the incorporation of the nanostructured material by using the inherent advantages of the CNTs character are only two. Electrodeposition and hydrothermal synthesis besides the chemical vapor deposition techniques can often be utilized to deposit CNTs down to the atom level, hence improving their contact with the electrolytes and share storage [54]. The marvelous performances of CNTs in conductivity can be enhanced via the hybrid formulation with metal oxides and other nanomaterials to realize novel high-performance electrode material with much enhanced capacitance performance (Fig. 3.4).

The mixture has been a good beginning since it has improved key performance indicators in quite an impressive manner and can be adopted as a point to carry out subsequent on researches on the establishment of sustainable and efficient energy storage systems. The advanced manufacturing techniques, the specific functionalization and the hybrid nanocomposites point to the never-ending perfection and evolution of the supercapacitor architecture in the shape of the CNTs and demonstrate their significant role in the creation of the supercapacitor of the next generation of the energy storage systems.

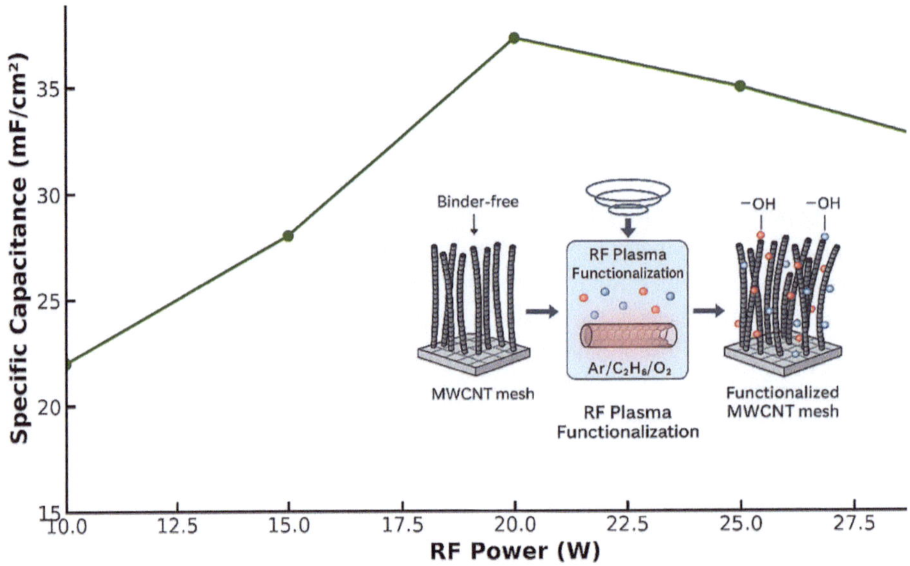

Fig. 3.4 Specific capacitance versus RF plasma power for MWCNT supercapacitor electrodes

3.6 Application of CNTs in Energy Storage

The world is in dire necessity of producing small, transportable yet strong-powered electronic gadgets. It proposes to study different types and characteristics of carbon nanotubes (CNTs) and comment on CNTs-based electronics with the emphasis on their benefits compared to conventional products. CNT based field emission devices which demonstrated remarkable features also are discussed. The current issues related to CNT-based electronics and the future potential of CNT in the electronic world are spoken about. This is, curiously enough, because of the long-standing tradition of the nation becoming independent, with myriads of fathers and mothers of the nation looking on [55]. Carbon nanotubes (CNTs) has potential application to the energy converter and storage technology and can be prepared through catalyzed expansion of expanded graphite at high temperature; however, further verification is needed on the mechanism of formation of carbon nanotubes in expanded graphite. Findings of this research incorporated the situ synthesis of spread carbon nanotubes within the distributed graphite (EG) via the catalytic pyrolysis procedure, using carbores P because the carbon source and $Co(NO_3)_3$•$6H_2O$ like the catalyst. Later on, carbon generated during the pyrolysis process was deposited on a Co particle surface resulting in the constant solid drying and recrystallization and eventually to the generation of EG carbon nanotubes [29] (Fig. 3.5).

Carbon nanotube can improve the efficiency of a number of energy storage systems (Table 3.6):

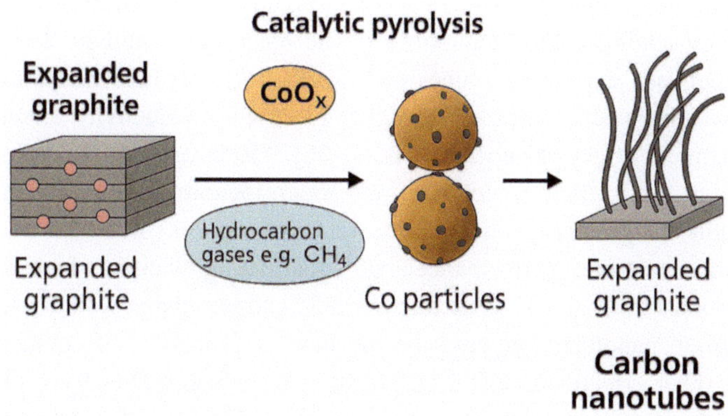

Fig. 3.5 Catalytic pyrolysis-based in-situ CNT growth in expanded graphite

Table 3.6 Roles of CNTs in energy storage

Storage system	Role of CNTs
LIBs	Anode material, conductive additive
Li–S batteries	Conductive scaffold for sulfur
Sodium-ion batteries	Mechanical buffer
Hybrid capacitors	High surface area electrode

- Carbon nanotubes are anodes and conductive addendums in Lithium-ion battery (LIBs);
- Li-sulfur batteries use carbon nanotube hosts for sulphur;
- Sodium-ion batteries A stabilized sodiation-desodiation reaction occurs with carbon nanotubes;
- Hybrid capacitors: The material used is carbon nanotubes as high-rate electrode materials.

LIB anode capacity:

$$C_{anode} = \frac{nF}{3,6M}$$

where:

- n—number of electrons exchanged;
- F—Faraday constant;
- M—molar mass of active material.

Supercapacitors have high power density and low charge discharge capacity which can be increased by adding carbon nanotubes in the electrode materials. There is a synergistic effect between carbon nanotubes and conducting polymers including polypyrrole which are used to make composites of higher electrical conductivity thereby increasing the electrochemical effect of supercapacitors [56]. Moreover, the structural qualities of carbon nanotubes, e.g. high aspect ratio and much-increased durability of composites, and lightweight energy storage. The ability to form networks-like structures provides a basis to enhance electrodes performance and maintain a lightweight which is necessary so that the electronics devices can be used in portable gadgets and powered vehicles. Studies show that carbon nanotubes increase the mechanical properties of composite materials and facilitate other effective types of charge transfer, which makes them fundamental in the creation of superior flexible electronics and construction of hybrid energy storage networks [57, 58].

Amorphous carbon nanotubes (a-CNTs) are gaining attention owing to their unique topological randomness and capability to expand electrochemical reaction sites, therefore, possibly to increase the capability of energy storage apparatus. The properties of the amorphous carbon nanotubes (a-CNT) have been simulated (including those with up to four walls) whose length is in the range of 200 to 3200 atoms ranging starting with gently random configurations which have been brought up by modelling through the Gaussian Approximation Potential. The machine learning random forest approach was used to predict critical characteristics, which includes density, height and diameter that will be needed in simulating the effective response of a-CNTs. The a-CNT models were of width 0.55 to 2 nm with average of 0.31 nm inter-wall spacing. The defect structure of a-CNTs was investigated and some new patterns were found. Electrical density of states and localisation in these phases were studied and delocalised electrons in the 3 space of the 132 system were identified as an important ingredient to inter-layer cohesisness. The geometrical mapping of conductance encourages axial translation instead of networked hexagons, whereas non-hexagonal parts of the network either block or distract electric transportation. It was studied how a low-frequency radial breathing mode appears, and the thermal conductivity at 300 K was estimated with the help of the Green–Kubo formula. Loading in energy storage matrices with a-CNTs increases their active content with the preservation of their structure, which, as a final product, leads to efficient and compact batteries and capacitors [59]. Some research currently being carried out concerning their electrochemical properties would indicate that the a-CNTs could yield improved performance characteristics, in addition to the necessity of having energy storage systems that are more effective and less voluminous.

CNTs have extend beyond batteries and super capacitors into enhancing energy conversion technologies, such as solar cells, fuel cells and so on. To provide the answer, the real-time time-dependent density functional tight-binding (TD-DFTB) approach in molecular dynamics modelling was used to clarify the mechanism of the formation of field emission centers in carbon nanostructure layers. By exposing them to laser beam, laser radiation will cause the formation of covalent bonds between nanotubes and graphene as well as forming dipole moment in the mixed nanostructures, thus, aligning it with the force lines of the radiation. All the samples were more than twofold harder after laser irradiation. The hybrid nanostructure (rGO(bl)-SWCNT) of a buffer layer of 3 nm thickness of rGO to a primary layer of 0.4 nm thick SWCNT attained a maximum hardness of 54.4 GPa. Moreover, the rGO adhesion to the substrate and electron transport between the substrate and rGO(bl)-SWCNT were more strengthened. The rGO(bl)-SWCNT cathode with the size of roughly 1 mm 2 revealed field emission value of 562 mA/cm^2 and held steady at 1 mA during 9 h. The fabrication technique of hybrid nanostructures has been found to apply in the manufacture of high performance, stable field emission cathodes besides its other uses that require nanomaterial coatings with high adhesion, strength and electrical conductivity. Still, another area is the hybridisation of carbon nanotubes with graphene and other materials, where there are opportunities associated with further increasing the absorption of light and charge carrier mobility. As transparent conductive electrodes on solar cells, CNTs improve the performance of the latter by providing a conductive path maximizing the effect of photons and decreasing energy loss [60, 61]. With the right incorporation, CNT/graphene hybrids research suggests can provide multi-functional energy storage capacity that switches between storage and production effortlessly, a vital requirement in the distributed energy system. To conduct the investigation, other topological and geometric parameters of a hybrid film are also varied, and they include nanotube diameter; the length of the graphene fragment along the zigzag direction (the Y-axis), e- the displacement of one graphene layer relative to another along the zigzag direction (the Y-axis), and the orientation (tilt angle) of bilayer graphene with respect to the nanotube surface. Sub- and nanometer (are often observed in the experimental setups), width chiral (6,3) and (12,8) SWCNTs were used to create atomistic models of graphene/SWCNT hybrid 2D structures. The band structure of the graphene/SWCNT (6,3) hybrid system displays an energy gap of approximately 0.1 eV, when layers of the same length of graphene are layered horizontally over the surface of the nanotube. In the hybrid nanostructure of the bilayer graphene sheets (having different dimensions in both zigzag direction and armchair direction) and SWCNT (12,8) when the angle between the sheets and the surface of the nanotube vary a certain electrical conductivity anisotropy is identified in the graphene/SWCNT (12,8) hybrid: in one case the electrical resistance is lower in the zigzag direction and higher in the direction of the hexagonal lattice of the graphene crystal. The anisotropy is due to different conduction channels number of zigzag and armchair directions. This structure of the bilayer graphene piece within the supercells of the examined graphene/SWCNT (12,8) hybrid films is an advantage of zigzag orientation. The bilayer of graphene has a zigzag nanoribbon structure that has its edge states, which can be located at the Fermi level [62].

Conclusively, carbon nanotubes form an important element in the development of energy storage technology. They also have outstanding features that enhance the performance of multiple technologies, such as at the level of lithium-ion batteries and supercapacitors. Their current significance in contemporary energy storage procedures is evidenced by the constant innovative solution that is found in recent studies concerning nanotechnology, which takes advantage of the unique qualities of carbon nanotubes. The further functionalization of chemicals and production of composite will likely increase the use of CNTs so that high performance sustainable energy storage systems can be created.

3.7 Application of CNTs Biosensing

CNTs have unique physico-chemical properties (large surface area, excellent electrical conductivity and flexibleness of surface functionalization potential) that would be very suitable to serve as platforms to support the development of complex biosensors. These characteristics have been extensively explored by the researchers and increased the sensitivity and specificity of various biosensing methods, thus, assisting in the achievement of significant advances in healthcare, environmental, and food safety.

SWCNTs have been a topic of interest in the fluorescence-based biosensing abilities due to the superior optoelectronic features. Biosensors can be viewed as powerful tools in current basic research and bio-diagnostics. The chemical sciences have a great role to play in their development. Light detecting sensor or probe e.g. fluorescence based offer a quicker, less invasive way of detecting analyte with simultaneously higher spatial/temporal resolution. Near-infrared (NIR) band has advantages in that it has reduced background and scattering in biological materials so it is useful in the tissue transparency window. Here, single-walled carbon nanotubes (SWCNTs) have shone as flexible near-infrared fluorescent bio-measurement parts. The current document is a summary of the progress with SWCNT-based near-infrared fluorescent molecular probes. We are also focused on methodologies of chemical design of the different analytes and aspects of photo physics and molecular recognition are also brought into discussion. Also, a broad variety of domains of application is considered, including chemical imaging of cellular systems, diagnostics, in vivo, and future perspectives. Photoemission characteristics of SWCNTs have been used in sensing biomolecules through light emission variations as a result of interacting with definite analytes or biological interaction making them the ideal biosensing materials [63, 64]. Molecular recognition Traditional optic biosensors operate largely by molecular recognition, whereby specific binding of ligands causes conformational changes that lead to optical perturbation of the emitting motif. Single-stranded DNA functions as the single walled carbon nanotubes (ssDNA-SWCNTs) have started to make a significant contribution to biological research through optical sensors. However, their functions have not been well understood with regard to the processes that govern them. This paper combines experimental and computational studies to prove that in this type of synthetic biosensors binding to ligands is not sufficient to achieve optical modulation. The chemical nature of the ligands has a considerable effect on the optical response when the ligand binds and this is similar to the events that occur in activity-based biosensors. We show that in ssDNA-SWCNT catecholamine sensors optical response is highly correlated to the electronic density of the aryl motif even in cases where the ligands have equally strong binding affinities. Notably, though strong correlating with electrochemical characteristics, we establish that catechol oxidation does not play a critical role in the generation of the optical response occurring at the sensor. We consider in which way these outcomes can serve as the basis to improve the work of the existing sensors and guide the creation of new biosensors of the type. Being sensitive to their immediate environment, these nanostructures allow the generation of highly selective biosensors that can be used to detect very small amount of target analytes as illustrated in the case of SWCNTs being used in the detection of proteins and nucleic acids synonymous to clinical diagnostic applications. CNT-based biosensors utilize:

- High surface-to-volume ratio
- Superior electrical properties
- Ability for surface functionalization.

Sensing mechanism:

Change in conductance:

$$\Delta G \propto \Delta n$$

where:

Δn is the change in carrier density upon binding of biomolecules (Table 3.7).

Applications:

- Glucose sensors
- DNA sensors
- Protein sensors
- Virus detection.

Biosensing Products (Examples, Electrochemical Biosensing Systems) An electrically sensitive potential application area of CNT is biosensing where the characteristics can significantly enhance the functionality of such a sensor. The work is a contribution of a novel,

Table 3.7 Properties of nanomaterials enabling efficient biosensing applications

Property	Biosensing benefit
Large surface area	High biomolecule loading
High conductivity	Sensitive signal transduction
Functionalizable surface	Selective detection
Biocompatibility	Suitable for in vivo applications

low-cost, fast, and cost-effective technique of defining an electrochemical transducing platform with GLEs conjugated to magnetic beads in biosensing. Such developed biosensors successfully determined S. typhimurium and L, monocytogenes label-free, indicating that it has the possibility to be applied to quantitatively detect pathogens in food safety analysis. The two infections had an experimental limit of detection (LoD) of 10 CFU/mL, meaning higher sensitivity of the suggested biosensors. The incorporation of carbon nanotubes into electrochemical biosensors leads to the improvement of the electron transfer kinetics to enable the detection of the biomolecules in a more sensitive way [65]. Thrombin diagnosis plays a paramount role in early detection of a number of diseases. Methods based on label-free, yet more importantly electrochemical aptamer sensors, are simple, specific, sensitive, and fast; however, issues concerning practical application exist. The purpose of this research was to define and to provide solutions to challenges associated with the application of label-free electrochemical aptamer-based sensors in the detection of thrombin to include high detection limits, biofouling, low sample incompatibility, inability to utilize whole blood, and low specificity. These problems were categorized based on the significant numbers of materials used in this research.

Different nanoparticles were employed with each of them having distinctive characteristics that help to solve some of the problems in biosensing. Compared with other types of nanoparticle, the nano-metallic objects offer an advantage in terms of charge transfer and they easily bind up with TBAs through self-assembly hence significantly enhance aptamer sensors sensitivity especially in the sandwich assay detection. Carbon nanoparticles have strong π-π interaction and a large number of functional groups, which gives them the possibility to be directly bound with the TBA and other nanomaterials, and make it easy to produce and increase the design flexibility of sensors. Enhanced control is offered by incorporating magnetic particles so that the reaction to the magnetic fields is rapid and redox active species could be transported to the electrode surface to measure the reaction in a sensitive and quick way. Among aptamer sensors is the use of polymer-based sensor that is known to be biocompatible and anti-biofouling and can be used to detect the presence of thrombin in whole blood analysis. Since none of the materials can satisfy all the requirements, as well as perform the best when a combination of different materials is used, the further development in this direction must be aimed at combining different functional nanomaterials to provide a synergistic effect, and consequently, increase the efficacy and electrocatalytic activity of the sensor. To this effect, carbon nanoparticles and porous nanomaterials can be used as the best solutions where other material components are anchored. The whole blood detection of thrombin is at the same time essential to the timely and effective diagnosis Moreover, the potentials of CNTs can be broadened through combining or coupling them with other nano-materials or bio-interfacing agents and producing hybrid biosensors. Investigators have designed multi-functional platforms which enhance electrochemical activity and separation and concentration of target biomolecules, by combining carbon nanotubes with magnetic beads or other nanocomposites. The study presents an original, low-cost, and rapid process of generating an electrochemical transducer that will be used in magnetic bead-based E. In the study, GLEs will be converted into an electrochemical transducer, and the subsequent use of magnetic beads will serve as bio sensing. All the electrode designs, including rhomboidal GLE and interdigitated structure, showed the excellent stability as well as repeatability and sensitivity with the minimum signal variation observed among numerous electrode sets. The engineered biosensors have successfully served the purpose of label-free detection of the presence of S. typhimurium and L. monocytogenes, and in this case, they demonstrate the ability to quantitatively detect the presence of the pathogens within food safety. Both of the infections had experimental limit of detection of 10 CFU/mL, which means the high sensitivity of the offered biosensors. Magnetic beads CNTs used in the biosensors might improve recognition component immobilization processes thereby significantly increasing response times and detection limits [66].

Food contamination with Listeria is currently identified within the molecular tests routinely done in diagnostic laboratories at now. There are no known phage-based test methods to recognize the L. monocytogenes in the production of food or the processing of food. We establish a potentially illustrative rapid diagnostic method with institutionalized electrochemical biosensor, that we use to detect and quantify the target of interest within the selective detection of bacteriophages used as bio-receptors to detect L. monocytogenes. Electrochemical biosensors are used as powerful substitutes to the molecular detection methods because of ease of use, enhanced specificity, sensitivity, and low cost. The exemplary attribute of bacteriophages in photosensor is that these bio recognizing elements are extremely stable in different environmental conditions and able to distinguish between the viable and non-viable bacterial cells. This impedimetric biosensing platform of detecting and analyzing the presence of L. monocytogenes was developed through immobilization of the P100 bacteriophage onto the modified carbon nanotubes quaternized polyethyleneimine utilizing a proprietary molecular tethering methodology. The resulting sensor showed increased selectivity and sensitivity to the target analyte, namely,'L. monocytogenes 0.84 CFU/ml. The biosensing platform is promising in its selectivity of target analyte but good selectivity, and shines through with its reliability. The recent investigation has revealed that carbon nanotubes-based impedimetric sensors could be effective in the detection of pathogens such as Listeria monocytogenes due to the immobilization of particular bioreceptors on surface-to-surface modified electrodes [67]. This approach has provided a faster and accurate technique to detect bacterial contamination, which reveals the practical applications of CNTs to the food safety area.

Another revolutionary aspect of CNT-based biosensors is that it is used in environmental monitoring. Electrochemical properties of CNTs are quite helpful in the detection of heavy metals and other pollutants in water supply. The vast nature of the environmental challenge and its correction can be demonstrated by the fact that carbon nanotubes (CNTs) with field-effect transistor (FET) biosensors have proven their effectiveness in this area and can be used to detect dangerous ions in real-time [68]. The application utilizes the sensitivity and the selectivity of the CNTs, and provides the capacity to detect

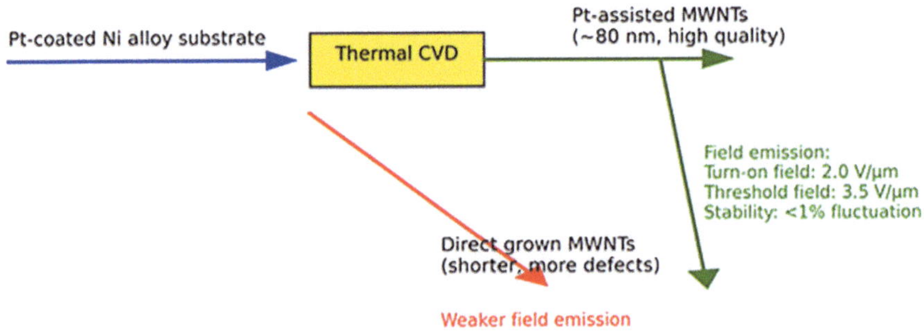

Fig. 3.6 Schematic of Pt-assisted MWCNTs field emitter synthesis and performance

and measure the contaminants at a site; this is also important to the protection of the environment. Thermal CVD synthetic field emitter made out of MWNTs has been observed on Pt-coated Ni alloy substrate and their structural and field emission Proprieties compared to instants of MWNTs. The Pt-catalyzed MWNTs were more extended and the diameters were larger about 80 nm and other structural qualities were better and their catalyst particle compositions were minimal. Pt-supplemented MWNTs proved all the above with a higher field emission of 2.0 v.mu m^{-1} and a turn on field of 3.5 v.mu m^{-1}. MWNTs demonstrated outstanding emission stability at such a current density level of approximately 30 mA/cm^2, where fluctuations of the current were less than 1%. This paper presented a new way of fabricating nano the field emitters carbon nanotubes to be utilized in electronic devices (Fig. 3.6).

The ongoing studies into abnormalities in knee cartilages though quite limited are quite promising. They are known to be hydrogel composite scaffolds that are more biocompatible and easily incorporated into the adjacent tissue, therefore, making them ideal in regenerating and repairing the chondrocyte. The use of hydrogel composite scaffold in implementation of knee joint cartilage defects may significantly reduce the pain level and restriction of movement in the patient. Moreover, hydrogel composite scaffolds consist of adjustable delivery properties. They can also produce beneficial cytokines and growth factors that elevate proliferation and differentiation of chondrocytes, and thus they accelerate the process of cartilage repair. Besides, these scaffolds can act as vehicles of transporting stem cells or genes into the cartilage defects, thus accelerating the cartilage healing process. However, there are some issues and restrictions that can be found in the modern research. The effectiveness of hydrogel composite Scaff resources over the long-term and their safety needs to be determined further. Second, more efficient and controllable production and manipulation of such scaffolds has some room to be developed. Third, clinical studies to reinforcing the efficacy of hydrogel composite scaffold in repairing different types and degree of cartilage lesions are necessary. Academics can focus on the following emerging trends in order to address such challenges. To enhance that mechanical properties, conducting studies on innovative composite material and cross-linking approaches could increase the strength and survivability of hydrogel composite scaffolds, so that they performed better under the mechanical pressures of the knee joint. To have a more controlled drug release, then advanced technology in drug-loading and release-control can be created using therapeutic agents and growth factors that will only release in the most appropriate time and dosage like using intelligent hydro-gels that response to certain physiological change (pH, temperature, or enzyme activity) to promote cartilage repair. The possibility of combining stem cell therapy to hydrogel composite scaffolds is quite high. In the process of directing stem cells to grow into chondrocytes and, therefore, enhance cartilage regeneration, one needs to be more efficient by carefully manipulating the microenvironment developed by the scaffolds. Introduction of signaling molecules such as extracellular matrix components to the scaffolds might be useful in chondrocytes differentiation of stem cells. Furthermore, the application of the novel technologies such as bioprinting will allow one to create a more accurate and customized scaffold depending on the unique set of needs of different patients and types of cartilage defects. This individual approach to the methodology will improve the outcomes of the treatment and patient satisfaction. The study of hydrogel composite scaffold in cartilage abnormalities of the knee portrays a positive prospect. Continuous technical progressive development together with the extensive research results in the concept of hydrogel composite scaffolds becoming an important therapeutic tool that can result in an improved patient patient management result in the presence of knee cartilage dysfunctions. Recently, carbon nanotubes were applied as conductive scaffolds in biosensors based on hydrogel with the capability of driving cell proliferation and providing electrical responses as the basis of the biosensing mechanism [69]. These developments indicate the multifaced use of CNTs in biomedical processes, as measuring elements and growth avocations of cells and tissue repair.

Drawing a conclusion, we can make sure that the carbon nanotubes is a versatile and powerful substance in biosensing, and the peculiar features can promote the development of various spheres. Carbon nanotubes make optical and electrochemical sensors sensitive and allow complicated biosensor formats, thus spearheading manufacture of the next generation of diagnostic and monitoring devices. With further investigations, the current study aimed at functionalization of CNTs and their integration with other nanomaterials is likely to deliver more advanced biosensing platforms that could address the current and pending challenges in healthcare, environmental surveillance and food safety.

3.8 Concluding Remarks

The use of carbon nanotubes (CNTs) is one of the most innovative and versatile nanotechnology materials, an outstanding composite of structural, electrical, mechanical, and thermal properties. Their unique and controllable properties and nanoscale structure have allowed their application in numerous ways in advanced electronic networks and energy storage systems, nanoelectron-mechanical devices (NEMS), field emitting devices, and biomedical technologies. CNTs in the energy storage architectures would help in better charge transport, greater electrode–electrolyte interface, and structural stability. These characteristics render them invaluable to the process of maximizing the operation of lithium-ion batteries, electrochemical capacitors and novel hybrid storage platforms. The current development in chemical functionalization and composite synthesis is predicted to improve the performance of energy storage devices that are based on CNTs so that high performance and sustainable developments, aligned with the international goals in energy efficiency and sustainability, can be developed.

- A number of technology applications are based on a complicated foundation on carbon nanotubes (CNTs): Field emission shows improved current density and lower turn-on voltages;
- Data storage using CNTs based memory devices exhibit scalability and fast speeds of operation;
- Their very sensitive and fast devices are made possible in NEMS by their remarkable electrical and mechanical properties;
- In energy storage and super capacitors carbon nanotubes (CNTs) enhance capacity, conductivity and stability;
- CNTs are also used within biosensing to offer high sensitivity detection, by means of functionalized surfaces;
- The existing research is directed at the mass-scale manufacture of carbon nanotubes of controlled features;
- Integration procedures of the devices manufacturing;
- Long term stability and biocompatibility in regions of medical purpose.

The continued evolution of CNT based technology will find a toll in a variety of applications in electronics, energy, healthcare and environmental surveillance. The application of CNTs, in field emission technologies, has been very promising in their ability to increase the performance of the electron source. The fact that the nature of the CNT structure and its emission characteristics has been still under investigation is likely to encourage further breakthroughs in the given sphere. CNTs demonstrate their flexibility in their application as the nanoelectromechanical systems (NEMS). The mechanical strength, light weight and excellent electrical properties that are inherent in them allow the production of high frequency resonators, switches, and sensors, into which there are potential medical applications in the area of diagnostics, environmental monitoring, and precision engineering. Designing tailored CNT structures/material hybrid would play a significant role in improving the functionalities of NEMS technologies. Carbon nanotubes (CNTs) are increasingly playing a very important role in biosensing activities. This application of CNTs in super capacitor technology evidences the importance of CNTs in increasing the storage level of energy. CNTs integration has brought significant changes in terms of capacitance, energy density, and cycling stability and has set the basis of sustainable and efficient energy storage systems. A fresh manufacturing process, surface modification, and hybrid nanocomposites have not stopped evolving towards an improvement in the performance of CNT-based supercapacitors. To sum up, CNTs have become the key in modern material science and nanotechnology. Their superior and modifiable qualities make them primary enablers of the technical advances in various areas. As synthesis improvements in accuracy, functionalizing approaches and integration strategies are developed, the full potential potential of CNTs will continuously become a reality and will deliver disruptive solutions in energy, electronics and biomedicine, and environmental technology. Detailed research involving CNTs requires that future be made as fruitful as possible by exploring the full potential of CNTs socially.

References

1. Yang, S., Li, N., Xiao, H., Wu, G., Liu, F., Pan, Q., Tang, L., Tan, X., & Yang, Q. (2022). Clearance pathways of near-infrared-ii contrast agents. Theranostics, 12(18), 7853–7883. https://doi.org/10.7150/thno.79209
2. Quintão, C., Camargo, L., Brandão, H., Saraiva, N., & Munk, M. (2022). Challenges in the use of nanostructures as carriers of nucleic acids in clinical practice. Einstein (São Paulo), 20. https://doi.org/10.31744/einstein_journal/2022rb5898
3. Said, Z., Sundar, L., Tiwari, A., Ali, H., Sheikholeslami, M., Bellos, E., & Babar, H. (2022). Recent advances on the fundamental physical phenomena behind stability, dynamic motion, thermophysical properties, heat transport, applications, and challenges of nanofluids. Physics Reports, 946, 1–94. https://doi.org/10.1016/j.physrep.2021.07.002
4. Reza, M., Afroze, S., Kuterbekov, K., Kabyshev, A., Bekmyrza, K., Haque, M., Islam, S.N., Hossain, M.A., Roy, H., Islam, M.S., Pervez, M.N. & Azad, A. (2023). Advanced applications of carbonaceous materials in sustainable water treatment, energy storage, and CO_2 capture: a comprehensive review. Sustainability, 15(11), 8815. https://doi.org/10.3390/su15118815
5. Shinde, M., Patwekar, M., Patwekar, F., Bajaber, M., Medikeri, A., Firdous, S.,Mukim, M., Soni, S., Mallick, J. & Jawaid, T. (2022). Nanomaterials: a potential hope for life sciences from bench to bedside. Journal of Nanomaterials, 2022(1). https://doi.org/10.1155/2022/5968131
6. Chakroborty, S., Pal, K., Nath, N., Singh, V., Barik, A., Soren, S., Panda, P., Asthana, N. & Kyzas, G. (2023). Sustainable synthesis of multifunctional nanomaterials from rice wastes: a comprehensive review. Environmental Science and Pollution Research, 30(42), 95039–95053. https://doi.org/10.1007/s11356-023-29235-9
7. Lee, J., Jung, H., & Park, H. (2023). Rare is beautiful? rareness, technology value, and the moderating role of search domain and knowledge maturity. Production and Operations Management, 32(4), 1019–1040. https://doi.org/10.1111/poms.13908
8. Rahman, K. (2023). Emerging nano-enable materials in the sports industry., 75–100. https://doi.org/10.21741/9781644902288-4
9. He, D., Xiao, X., Hu, G., Zhang, W., Yu, G., Liu, Y., Lin, Y., Lin, H., Li, X., Diao, Y., Tang, Y. & Li, H. (2024). Nitrogen-doped multiwalled carbon nanotubes trigger immune responses and inhibit fat deposition. Advanced Materials Interfaces, 11(36). https://doi.org/10.1002/admi.202400007

10. Amaro, S., Maciel, C., Rodrigues, J., Freitas, A., Vasconcelos, L., Barros, A. & Ferreira, M. (2023). Investigation of the synergistic effect of layer-by-layer films of carbon nanotubes and polypyrrole on a flexible electrochemical device for paraquat sensing. Chemosensors, 11(8), 420. https://doi.org/10.3390/chemosensors11080420
11. Sen, S. and Manik, N. (2023). Estimation and modification of electrical parameters of organic device in the presence of single walled carbon nanotubes. https://doi.org/10.5772/intechopen.107728
12. Purusothaman, M., Venkatesh, A., Jotheeswaran, S., & Kumar, D. (2023). Review on emission reduction of VCR system using various nano-refrigerants. Environmental Quality Management, 33(4), 579–593. https://doi.org/10.1002/tqem.22119
13. Lal, S., Bhat, V., & Sharma, S. (2023). Computational analysis of nm-polynomial based topological indices and graph-entropies of carbon nanotube y-junctions. https://doi.org/10.48550/arxiv.2301.02169
14. Rashko, M., Othman, M., & Hamad, A. (2022). Effects of doping cadmium atoms on the electronic and optical properties of (n, 0) ZigZag SWCNTs: dft approach. ECS Journal of Solid-State Science and Technology, 11(8), 081009. https://doi.org/10.1149/2162-8777/ac861b
15. Gabellone, S., Vanni, S., Fausti, V., Miserocchi, G., Liverani, C., Spadazzi, C., Cocchi, C., Calabrese, Ch., Cavaliere, D., Pacilio, C.A., Ercolani, G., Pieri, F., Gurrieri, L., Riva, N., Jones, R. & Vita, A. (2024). Exploring nanotechnology solutions for improved outcomes in gastrointestinal stromal tumors. Heliyon, 10(23), e40596. https://doi.org/10.1016/j.heliyon.2024.e40596
16. Ayub, H., Khan, L., McCarthy, É., Ahad, I., Fleischer, K., & Brabazon, D. (2023). Investigation on optical absorption and reflection of carbon nanotubes mixed copper composites for laser sintering process improvement. Metals, 13(12), 1984. https://doi.org/10.3390/met13121984
17. Nasture, A., Ionete, E., Lungu, F., Spiridon, S., & Pătularu, L. (2022). Water quality carbon nanotube-based sensors technological barriers and late research trends: a bibliometric analysis. Chemosensors, 10(5), 161. https://doi.org/10.3390/chemosensors10050161
18. Shin, J., Kovaleski, S., Maschmann, M., Gunter, C., & Lindsay, M. (2025). Precise 3d CNT array catalyst patterning using two-photon lithography for enhanced liftoff and scalability., 67. https://doi.org/10.1117/12.3051129
19. Zheng, Y., Liu, Y., Wang, L., Liu, J., Wei, J., Chen, L., An, K., Zhang, X., Ye, H., Zhou, H., Tao, H., Yin, Y., Ouyang, X. & Li, C. (2022). Vertically aligned boron-doped diamond hollow nanoneedle arrays for enhanced field emission. ACS Applied Nano Materials, 5(8), 10878–10889. https://doi.org/10.1021/acsanm.2c02180
20. Yao, Q., Wu, Y., Song, G., Xu, Z., Ke, Y., Zhan, R., Chen, J., Zhang, Y. & Deng, S. (2024). Effect of crystallinity on the field emission characteristics of carbon nanotube grown on w-co bimetallic catalyst. Nanomaterials, 14(10), 819. https://doi.org/10.3390/nano14100819
21. Karmakar, S., Sarkar, P., Mistari, C., Trivedi, R., Chakraborty, B., More, M., Droopad, R. & Haque, A. (2023). Tubular diamond as an efficient electron field emitter. Acs Applied Electronic Materials, 5(7), 3592–3602. https://doi.org/10.1021/acsaelm.3c00317
22. Tang, H., Liu, R., Huang, W., Zhu, W., Qian, W., & Dong, C. (2022). Field emission of multi-walled carbon nanotubes from pt-assisted chemical vapor deposition. Nanomaterials, 12(3), 575. https://doi.org/10.3390/nano12030575
23. Nwanno, C., Thapa, A., Watt, J., Bendayan, D., & Li, W. (2024). Field emission properties of cu-filled vertically aligned carbon nanotubes grown directly on thin cu foils. Nanomaterials, 14(11), 988. https://doi.org/10.3390/nano14110988
24. Hussain, S., Akbar, U., Siddique, M., Ibrar, M., Ali, Z., Waqas, M., Ahmad, Sh. A., Abbas, S. M., Ullah, H. & Anwar, S. (2023). Synthesis, properties, and applications of carbon nanotubes: an overview. Scientific Inquiry and Review, 7(3), 95–124. https://doi.org/10.32350/sir.73.07

25. Adhikari, B., Bhotkar, K., Kim, J., Yoo, S., Choi, E., & Park, K. (2022). Beam trajectory analysis of vertically aligned carbon nanotube emitters with a microchannel plate. Nanomaterials, 12(23), 4313. https://doi.org/10.3390/nano12234313
26. Lee, D., Kim, S., Hong, S., Madrona, C., Oh, Y., Park, M., Komatsu, N., Taylor, L. W., Chung, B., Kim, J,m Hwang, J. Y., Yu, J., Lee, D. S., Jeong, H. S., You, N. H., Kim, N. D.m Kim, D. Y., Lee, H. S., Lee, K. H., Kono, J., Wehmeyer, G., Pasquali, M., Vilatela, J. J., Ryu, S. & Ku, B. (2022). Ultrahigh strength, modulus, and conductivity of graphitic fibers by macromolecular coalescence. Science Advances, 8(16). https://doi.org/10.1126/sciadv.abn0939
27. Bahrami, M. and Feli, S. (2023). Synergistic influence of mwcnts/rgo on low-velocity impact response and mechanical properties of carbon fiber/epoxy composite. Proceedings of the Institution of Mechanical Engineers Part N Journal of Nanomaterials Nanoengineering and Nanosystems. https://doi.org/10.1177/23977914231176865
28. Malviya, K., Naganagowda, H., Pavithra, K., Varma, S., Raju, C., & Sharma, R. (2024). Heat transfer analysis of mhd oscillatory swcnt/mwcnt-H_2O hybrid nanofluid flow in a channel. Zamm—Journal of Applied Mathematics and Mechanics / Zeitschrift Für Angewandte Mathematik Und Mechanik, 104(4). https://doi.org/10.1002/zamm.202300366
29. Wang, Q., Yuan, H., Zhang, M., Yang, N., Cong, S., Zhao, H., Wang, X., Xiong, Sh., Li, K. & Zhou, A. (2023). A highly conductive and supercapacitive MXene/n-CNT electrode material derived from a MXene-co-melamine precursor. Acs Applied Electronic Materials, 5(5), 2506–2517. https://doi.org/10.1021/acsaelm.2c01768
30. Ramar, V. and Ambedkar, B. (2022). Critical review on carbon-based nanomaterial for carbon capture: technical challenges, opportunities, and future perspectives. Energy & Fuels, 36(22), 13479–13505. https://doi.org/10.1021/acs.energyfuels.2c02585
31. Awasthi, S., De, S., & Pandey, S. (2024). Electrodeposited carbon nanostructured nickel composite coatings: a review. Heliyon, 10(8), e26051. https://doi.org/10.1016/j.heliyon.2024.e26051
32. Jain, P., Jha, S., & Ingole, P. (2022). Strong metal–support interaction in copper hexacyanoferrate nanocube decorated functionalized multiwall carbon nanotubes for enhanced bi-functional oxygen electrocatalytic activity and stability. Sustainable Energy & Fuels, 6(4), 1094–1107. https://doi.org/10.1039/d1se01635k
33. Nugroho, A., Nursanto, E., Curie, C., Oktaviano, H., Ainurrachma, F., & Trisunaryanti, W. (2022). Utilisation of gelatin as nitrogen source for n-doped carbon nanotubes and its performance for the oxygen reduction reaction. Advances in Natural Sciences Nanoscience and Nanotechnology, 13(3), 035004. https://doi.org/10.1088/2043-6262/ac8660
34. Zhao, J., Chen, Y., An, Z., Zhang, M., Wang, W., Guo, Q., Li, Y., Han, Sh. & Zhang, L. (2025). Toward the next generation of sustainable aluminum-ion batteries: a review. Green Chemistry, 27(2), 352–378. https://doi.org/10.1039/d4gc04505j
35. Malyadri, T., Kumar, J., & Nagamadhu, M. (2023). Influence of CNT particle on mechanical properties of epoxy composites. Journal of Mines Metals and Fuels, 2229–2236. https://doi.org/10.18311/jmmf/2023/36254
36. Okolie, O., Kumar, A., Edwards, C., Lawton, L., Oke, A., McDonald, S., Thakur, V. K. & Njuguna, J. (2023). Bio-based sustainable polymers and materials: from processing to biodegradation. Journal of Composites Science, 7(6), 213. https://doi.org/10.3390/jcs7060213
37. Song, Q., Gao, H., Cheng, L., Xiao, Z., Li, D., Wang, Y., Xie, M., Filler, N. A. & Zhu, M. (2025). Surface nanostructure fabrication by initiated chemical vapor deposition and its combined technologies. ACS Macro Letters, 14(2), 214–224. https://doi.org/10.1021/acsmacrolett.4c00793

38. Torkashvand, Z., Shayeganfar, F., & Ramazani, A. (2024). Nanomaterials based micro/nanoelectromechanical system (MEMS and NEMS) devices. Micromachines, 15(2), 175. https://doi.org/10.3390/mi15020175
39. Erdogan, R., Alkhaled, M., Kaynak, B., Alhmoud, H., Pisheh, H., Kelleci, M., Karakurt, I., Yanik, C., Şen, Z. B., Sari, B., Yagci, A. M., Özkul, A. & Hanay, M. (2022). Atmospheric pressure mass spectrometry of single viruses and nanoparticles by nanoelectromechanical systems. ACS Nano, 16(3), 3821–3833. https://doi.org/10.1021/acsnano.1c08423
40. Mu, W. (2025). Adhesive force between biconcave red blood cell membrane and bulk substrate. Membranes, 15(3), 89. https://doi.org/10.3390/membranes15030089
41. Nguyen, S., Montesinos-Castellanos, A., Madou, M., Martínez-Chapa, S., Ornelas-Soto, N., Flores, C., & Salazar, A. (2024). Characterizing joule heating effects on single suspended carbon fibers: calculating the temperature coefficient of resistance for localized carbon nanotube growth. The Journal of Physical Chemistry C, 128(30), 12593–12603. https://doi.org/10.1021/acs.jpcc.4c01116
42. Bognash, M. and Asokanthan, S. (2022). Bouncing dynamics of electrostatically actuated nem switches. Nano Express, 3(2), 025004. https://doi.org/10.1088/2632-959x/ac4668
43. Abouelregal, A. and Tiwari, R. (2023). Computational analysis of thermoelastic vibrations of functionally graded nonlocal nanobeam excited by thermal shock. Journal of Vibration and Control, 30(13-14), 3105–3116. https://doi.org/10.1177/10775463231190263
44. Advincula, P., Beckham, J., Choi, C., Chen, W., Han, Y., Kosynkin, D., Lathem, A., Mayoral, A., YAcaman, M. J. & Tour, J. (2023). Tunable hybridized morphologies obtained through flash joule heating of carbon nanotubes. ACS Nano, 17(3), 2506–2516. https://doi.org/10.1021/acsnano.2c10125
45. Dimitrov, B. and Konaklieva, S. (2023). A battery cell equalisation system based on a supercapacitors tank and dc–dc converters for automotive applications. World Electric Vehicle Journal, 14(7), 185. https://doi.org/10.3390/wevj14070185
46. Liu, J., Chen, S., Mao, N., Cheraghi, E., Sun, Y., Wang, J. & Yeow, J. (2022). Computational simulation and phenomenological analysis of freestanding carbon nanotube coulomb explosion induced by field emission. Journal of Materials Chemistry C, 10(42), 15872–15882. https://doi.org/10.1039/d2tc02275c
47. Liu, J., Wang, J., Cheraghi, E., Chen, S., Sun, Y., & Yeow, J. (2022). Improvement of field emission performances by dmso and pedot:pss treated freestanding CNT clusters. Nanoscale, 14(41), 15364–15372. https://doi.org/10.1039/d2nr04205c
48. Liang, R., Du, Y., Lin, J., Chen, J., & Xiao, P. (2022). Facile-synthesized Ni-metal–organic framework/nano carbon electrode material for high-performance supercapacitors. Energy & Fuels, 36(13), 7115–7120. https://doi.org/10.1021/acs.energyfuels.2c01058
49. Yang, W. and Zhitomirsky, I. (2022). Colloidal processing of Mn_3O_4-carbon nanotube nanocomposite electrodes for supercapacitors. Nanomaterials, 12(5), 803. https://doi.org/10.3390/nano12050803
50. Pour, G., Ashourifar, H., Aval, L., & Solaymani, S. (2023). CNTs-supercapacitors: a review of electrode nanocomposites based on CNTs, graphene, metals, and polymers. Symmetry, 15(6), 1179. https://doi.org/10.3390/sym15061179
51. Pajootan, E., Ye, M., Zhang, M., Niroumandrad, S., Omanovic, S., & Coulombe, S. (2022). Plasma-functionalized multi-walled carbon nanotubes directly grown on stainless steel meshes as supercapacitor electrodes. Journal of Physics D Applied Physics, 55(19), 194001. https://doi.org/10.1088/1361-6463/ac4fd8
52. Lv, K., Zhang, J., Zhao, X., Kong, N., Tao, J., & Zhou, J. (2022). Understanding the effect of pore size on electrochemical capacitive performance of MXene foams. Small, 18(27). https://doi.org/10.1002/smll.202202203

53. Zhang, W., Miao, J., Zuo, X., Zhang, X., & Qu, L. (2022). Weaving a magnificent world: 1d fibrous electrodes and devices for stretchable and wearable electronics. Journal of Materials Chemistry C, 10(38), 14027–14052. https://doi.org/10.1039/d2tc02524h
54. Deng, Y., Zhao, Y., Peng, K., & Yu, L. (2022). One-step hydrothermal synthesis of MoO_2/MoS_2 nanocomposites as high-performance electrode material for supercapacitors. ACS Applied Materials & Interfaces, 14(44), 49909–49918. https://doi.org/10.1021/acsami.2c11244
55. Maheswaran, R. and Prabu, S. (2022). A critical review of the role of carbon nanotubes in the progress of next-generation electronic applications. Journal of Electronic Materials, 51(6), 2786–2800. https://doi.org/10.1007/s11664-022-09516-8
56. Wang, B. (2024). Preparation of carbon nanotube transistors and their applications in integrated circuits. Highlights in Science Engineering and Technology, 111, 144–147. https://doi.org/10.54097/b45htv75
57. Yue, X., Wang, G., Wang, J., Fan, L., Hao, J., Wang, S., Yang, M. & Liu, Y. (2022). Electrochromic and electrochemical properties of Co_3O_4 nanosheets prepared by hydrothermal method. Coatings, 12(11), 1682. https://doi.org/10.3390/coatings12111682
58. Pyo, S., Eun, Y., Sim, J., Kim, K., & Choi, J. (2022). Carbon nanotube-graphene hybrids for soft electronics, sensors, and actuators. Micro and Nano Systems Letters, 10(1). https://doi.org/10.1186/s40486-022-00151-w
59. Ugwumadu, C., Thapa, R., Al-Majali, Y., Trembly, J., & Drabold, D. (2023). Formation of amorphous carbon multi-walled nanotubes from random initial configurations. Physica Status Solidi (B), 260(3). https://doi.org/10.1002/pssb.202200527
60. Gerasimenko, A., Kuksin, A., Shaman, Y., Kitsyuk, E., Fedorova, Y., Murashko, D., Shamanaev, A. A., Eganova, E. M., Sysa, A. V., Savelyev, M. S., Telyshev, D. V., Pavlov, A. A. & Glukhova, O. (2022). Hybrid carbon nanotubes–graphene nanostructures: modeling, formation, characterization. Nanomaterials, 12(16), 2812. https://doi.org/10.3390/nano12162812
61. Pander, A., Onishi, T., Hatta, A., & Furuta, H. (2022). Fabrication of self-assembling carbon nanotube forest fishnet metamaterials. Nanomaterials, 12(3), 464. https://doi.org/10.3390/nano12030464
62. Slepchenkov, M., Barkov, P., & Glukhova, O. (2023). Electronic and electrical properties of island-type hybrid structures based on bi-layer graphene and chiral nanotubes: Predictive analysis by quantum simulation methods. *Coatings, 13*(5), 966. https://doi.org/10.3390/coatings13050966
63. Ackermann, J., Metternich, J., Herbertz, S., & Kruss, S. (2022). Biosensing with fluorescent carbon nanotubes. Angewandte Chemie, 61(18). https://doi.org/10.1002/anie.202112372
64. Krasley, A., Chakraborty, S., Vuković, L., & Beyene, A. (2025). Molecular determinants of optical modulation in ssdna–carbon nanotube biosensors. Acs Nano, 19(8), 7804–7820. https://doi.org/10.1021/acsnano.4c13814
65. Govedarica, M., Milosevic, I., Jankovic, V., Mitrovic, R., Kundacina, I., Nastasijevic, I., & Radonic, V. (2025). A Cost-Effective and Rapid Manufacturing Approach for Electrochemical Transducers with Magnetic Beads for Biosensing. *Micromachines, 16*(3), 343. https://doi.org/10.3390/mi16030343
66. Yousef, H., Liu, Y., & Zheng, L. (2022). Nanomaterial-based label-free electrochemical aptasensors for the detection of thrombin. Biosensors, 12(4), 253. https://doi.org/10.3390/bios12040253
67. Zolti, O., Suganthan, B., Maynard, R., Asadi, H., Locklin, J., & Ramasamy, R. (2022). Electrochemical biosensor for rapid detection of listeria monocytogenes. Journal of the Electrochemical Society, 169(6), 067510. https://doi.org/10.1149/1945-7111/ac7a63

68. Yang, X., Du, J., Chen, H., Cui, C., Liu, H., & Zhang, X. (2024). Advances in field-effect tube biosensors constructed on silicon nanowires, graphene and carbon nanotubes for heavy metal ion detection. Sensor Review, 44(5), 525–538. https://doi.org/10.1108/sr-03-2024-0184
69. Tang, D., Song, H., Yan, C., Luo, Y., Su, X., & Ruan, S. (2025). Hydrogel composite scaffold repairs knee cartilage defects: a systematic review. RSC Advances, 15(13), 10337–10364. https://doi.org/10.1039/d5ra01031d

4 Future Outlook

The trend of application of carbon nanotubes (CNTs) relates closely to their peculiar physical chemical properties where it can take pivotal positions in various fields. The carbon nanotubes (CNTs) are supposed to be very crucial in energy storage applications due to its large surface area and electrical conductivity, which are important in devices such as super capacitors. According to various researches, CNTs use as fillers can enhance the functioning of such devices. Recognizing their competence to extend supercapacitors efficiency, Sumana et al. claim that high-surface materials like CNTs might augment the effectiveness of biodegradable films used as supercapacitors. What is more, although there is no direct reference to energy harvesting made in relation to specific Vishwanath, his work was explored in relation to CNTs features that may be beneficial in the energy consumption [1, 2]. This seeks to establish CNTs in the context of sustainable and renewable energy solutions which is a major contributor to the future of energy storage devices.

This will affect integration of the CNT in a variety of commercial applications with support of the continued advancement of the CNT research; the synthesis procedure. They used and placed acetone ($(CH_3)_2CO$) which was thermally decomposed carbonaceous source in a primary argon gas flow. The experimental study carried out with two different ratios of precursor gas of $((CH_3)_2CO + Ar)/Ar$ and two different morphologies of carbon layers were taken after one hour of deposition. The film composition at the ratio of 1/5 was not uniform and the regions intensified with SiO_2 occupied approximately 5% of the surface of the substrate to produce a sparkly rugged look. With Raman spectroscopy, these components were found to include the diamond like carbon structures. This when compared to a background deposition at a ratio of 1/0 led to forming a homogenous film which stained deep brown to black. According to Raman spectroscopy, this layer was

R. G. O. Abaszade and A. K. Haghi, *Carbon Nanotube Devices for Nanoelectronics*, Synthesis Lectures on Solid State Materials and Devices,
https://doi.org/10.1007/978-3-032-02114-4_4

found to be multilayered and graphene-high density structural impurities. Further structural characterization conditions identified the aggregate formation of diamond and that of its polytypes, intermittent crystallization of SiC nanocrystals and entanglement of SiC/graphene-like aggregate domains, which grow more prominent in the relative unusual situations during typical deposition ones. The change of carbon hybridization provided us with the perspective of transitions in the carbon hybridization enabled by complementary molecular dynamics simulations. The simulations have shown to be true that, under the conditions of interest, the direct conversion of sp^3- to sp^2-hybridized carbon is not possible. However, changes in intra-hybridizations e.g. sp^2 to sp^2 and sp^3 to sp^3 could be found to change freely in a broad variety of thermodynamic conditions. The use of CVD method is the most common method of depositing CNTs and the constant efforts put in to perfect this mode of production will create a scalable method of CNT production with a way of halting the cost of doing so thus increasing the number of readily available CNT products to industry [3]. Nevertheless, despite the possible difficulties, one must confess that a person must know about the issues that are going to be encountered. When introducing CNTs in different systems, the biocompatibility of the material, its stability and its environmental effect needs to be considered with large significance. Their toxicology deserves closer examination so that it is unequivocal about its safety in the biomedical arena due to the examination of Ushakov et al. the emphasis of whose work concerns the nature of materials required in its sustainable use [4]. The conscious progress of CNT-based technology will be developed by addressing them with sober thinking considering the advantages over possible risks.

The future of carbon nanotube promises much in terms of various aspects such as energy, environmental technology, electronic and health care. Ongoing nature of multi-disciplinary research will thus become an inevitable part of the solution to the problems of the applications of CNTs in the effort of us to capitalize on the potentials of the habits using them exquisitely.

4.1 Advances in Synthesis and Purification

Purification and production of carbon nanotubes (CNTs) improved during the last several years due to the necessity to utilize high quality material with specific characteristics in the majority of applications. The response provides an overview of published material and methodology in the area with references to significant changes in technologies of synthesis and purification processes.

It is easy to reach CNT synthesis utilization of Hot-wall CVD because the catalyst composition and temperature can be controlled properly to guarantee controlled growth. Another aspect that is pointed out in the present researches is that the composition and properties of the catalysts are major factors that determine the final morphology and crystallinity of CNTs. The high-quality lignin based CNTs was prepared by a floating catalyst CVD technique with the use of iron catalyst in which the primary carbon source in the synthesis process was carbon monoxide. Duraia et al. performed a single step of green variety of synthesis reaction to display the major importance of the contact of particles of catalysts with carbon feedstock in the creation of the many types of morphologies of CNTs such as the trumpet formed types. These findings give emphasis on the fact that the use of carbon source and the design of the catalyst has a direct effect on the productivity and quality of CNT production. Bati and co-workers have proposed a mechanochemical pathway to the synthesis of corannulene, and it possible that similar methods can be refined to synthesize carbon nanotubes, which would lead to low-cost scalable commercial methods of production [5–7].

These developments give reason as to why new methods to synthesis have to be further considered in the effort to increase the yield, as well as, the structural characteristics of carbon nanotubes (CNTs). This paper suggests a multi-step purification treatment which was performed sequentially to reduction with steam and with hydrochloric acid that allows reduction of iron content in single-walled carbon nanotubes (SWCNT) sample to a level as low as 0.2 wt.%. Researches on Raman spectroscopy have made it known that there will be no structural defects with the backbone of the SWCNTs due to this purification process. We are also going to provide an effective procedure of quantitative analysis of the iron through UV-Vis spectroscopy as an end to research. The post combustion residue is dissolved in HCl to produce hexaaquairon (III) chloride so that there is iron analysis using UV-Vis. The method is characteristically favorable since it does not rely on any external complexing molecules. Additionally, Martincic and Tobias have denoted the utility of steam treatment in terms of purification of highly oriented CNTs arrays consequently enhancing the corresponding crystallinity and demising the degree of flaws [8]. All these purification techniques are very essential in checking whether CNTs can be able to meet those high standards that one should meet in order to be used in new high-tech applications and most importantly electronics industry and nanocomposites. The modern combination of the various synthesis methods and impactful purification procedures is approaching to aid advancement of high-performance carbon nanotubes (CNTs) that may have application in many regions including electronics and green clean-up.

4.2 Integration in Next-Generation Electronics

Modern materials and innovative technology are a modern technique, which would be needed in order to create the next generation electronics. This increase depends on the design of electronic objects in the direction of multitasking, flexibility, and optimal performance. Other studies aroused recently concern themselves with multiple techniques and materials that facilitate the such integration working on issues like heat control, electromagnetic interference (EMI) and component miniaturization. One of the principal goals is formulation of the multifunctional material, which is able to control thermal properties and has electromagnetic interference shielding capacity. Due to the trend of the integration of electronic systems and the high density of power, 5G and renewable energy are emerging very rapidly and have a growing need of materials with great flexibility, mechanical performances, and multi-functional capabilities. A ribbed polyamide commercial mesh is used on the bottom side to increase the mechanical characteristics. The polyamide fabric is treated with the electroless nickel process in order to deposit a thick layer of nickel to the polyamide fabric with higher electromagnetic interference shielding effectiveness. The nickel-plated polyamide fabric also getting the WPU coating to serve the same purpose, to guard the cladding as well as to adhesive the top picture and bottom together. Experimental simulations show the realism of the concepts portrayed in the movie to find a solution to the issue of heat-handling and even electrical shielding. The MXene based compositions have been encouraging particularly with the aim of improving on electromagnetic shielding and thermal management qualities that are just paramount to the wearable electronics that emit at least a lot of heat and subjected to electromagnetic interferences [9].

Also, one can remember about the dominance of two-dimensional (2D) materials including graphene and the transitional metal dichalcogenide whose electrical properties are outstanding and the integration density is high. They can arrange the materials in order to have ambipolar transport properties that are significant in the deep electronic circuit merchandise. The ability to uniformly manipulate the charge carrier statistics - i.e. charge type and density - is a bridge to unlocking the bottleneck of pent-up potential of the 2D ambipolar semiconductors in next generation, high performance electronic and integrated circuit devices. GeH possesses commendable band gap, ambipolar cloth and high carrier mobility because of which it experiences lateral polarization amid local bias of emitting (opposing) of pm8V that has a variance in work of 0.6 eV. Such a defined control of charge carrier properties in hydrogenated germanene (GeH) is possible by heterointerface engineering a heterointerface ferroelectric material (PMN-PT) by which the control of type and carrier concentration by electrostatic charge carrier doping is achieved with a fine degree of control over the type and concentration of carriers. It is rather a plus that they can be scaled appropriately, and the materials can be applied in the large devices and flexible electronics [10, 11]. It is also researching, new production of methods, which will enable production of such advanced materials to work in real devices along with advances in material. Flexible transparent conductive electrodes (TCEs) in large area and uniformity and excellent optoelectronic properties that can be incorporated in transparent wearable electronics, such as antennas, sensors, and supercapacitors remain a challenge and key to enabling optoelectronic applications. By use of printing and coating, we demonstrate that conductive ink solution processing could lead to generation of transparent conductive electrodes with outstanding uniformity and low cost, which offers an untapped opportunity in largescale commercial production of transparent electronics. However, it is an enormous task to come up with such conductive inks and create uninterrupted conductive films without such a percolation problem. Here, we describe, to our knowledge, the first scalable patternable $Ti_3C_2T_x$ MXene aqueous ink slot-die coating process which allows the uniform deposition of transparent conductive electrodes (TCEs) on large-area substrates with exceptional optoelectronic performance, an average direct current conduction of 13,000 ± 500 S cm^{-1}. Further, the ink compositions can be optimised such that, the conductive MXene films can shift easily between transparent and opaque not showing much roughness and shine. Well defined as-prepared slot-die-coated MXene TCEs have admirable charge retaining properties when utilised in the framework of supercapacitor implementation. Conductive films made using MXene inks have also been made smooth using the scalable slot-die coating to a device that is significant in transparent fields and EMI shielding. The 2D materials are also famous in terms of the advantages of short-channel effects along with power consumption in field-effect transistors (FETs). We computational examine the electrical and ballistic transport properties of monolayer Be2C that consists of quasi-planar hexacoordinated carbons by applying the non-equilibrium Green-function scheme which was revealed on density functional theory. The Be2C is of monolayer bonded with planar anticonventional and possesses a gap of direct band 1.53 eV. The p-type Be_2C HP FET has its high ion value that is 2767 1A fm^{-1} [12, 13]. In that same way the advance of the stretchy solder methods with the liquid metals also enable the integration of the rigid bits and pieces with the flexy substrates, thus the more functional electronic systems. Next-generation electronics A notable trend when it comes to next-generation electronics is the introduction of bioelectronics e.g. wearable devices. The improvement of the user experience and operability, the creation of skin-like electronics that can be effortlessly placed on the skin is adopted to make the product more effective [14, 15]. The gross depletion of the conventional source of power results in hasty studies towards generation of self-powered electrical appliances in order to advance more ahead in the field of technology in electronics. Moisture-electric generators (MEGs) are becoming a viable and possible technology to abolish the energy concept as the world goes by it uses the ambient humidity which is freely available in huge amounts and works with wearable gadgets, so there is no necessity to use any form of outer energy. In recent times MEGs have been dominated by inorganic, metallic or petroleum-based polymeric materials as active material constituents and they may have adequate current capacity, but not adequate flexibility and stretch capabilities to enable body-borne electronic applications. The elastomer-based ionic hydrogel active part of MEG was processed to the same thickness as the active layer, on which the formulation of the elastic-based active layer of an ionic hydrogel active part was performed with the ionic hydrogel casted on a PET sheet, and the copper tape was bonded to the both surface of ionic hydrogel. All of the above-mentioned hydrogels were highly optimized and studied for spectroscopy analysis, swelling, water retention, mechanical and rheology measurements. The mechanism of MEG actuation is originated from the combined effect of ion and redox reaction between cathode and anode. This is an independent electronic sensor device in addition to operating on the other generation level in order to monitor the other physical activities depending on the breath patterns. This gadget is also usable to track hands. To guarantee the survival and functionality of the gadgets, such devices pose advanced materials that should be biocompatible, soft as well as possess such characteristics as self-healing and energy harvesting [16].

This study in the realm of hybrid systems that combine the electrical and photonic capabilities is accelerating. Higher systems come in the form of electro photonic (EPh) circuits and this allows the control of photons and electrons run concurrently in the same integrated circuit that leads to the invention of special sensors that could be applied in various applications. An EPh biotin detections circuit compatible with CMOS is proposed and a silicon based light emitting source is monolithically combined. These incorporated features are inbuilt light and a waveguide having a p-n photodiode which are made to monolithic chip. The waveguide was investigated in the fashion of being functionalized at the surface so that it could be biotinylated in the future so that it could be used in biosensing. AFM, optical microscopy, Raman spectroscopy and photoluminescence served to scan the modified surfaces. The differences were surface alterations of the waveguide differences of photocurrent intensity measurements that were observed in the photodiode, the photocurrent intensity of the surface alterations owing to functionalization and biotinylating as an indication that the EPh circuit may be used as a biosensor. Electro photonic circuits that integrate engineering electrical and optical systems are to transform the field of communications due to their capability to transmit data faster along with processing [17]. All these integrations are caused by the advancements in the materials science, specifically in the manufacturing of low-power and high-performance semiconductors that could successfully work in such a hybrid system. Low-power consuming (ferroelectric) device concepts are currently being developed, compatible with established complementary metal–oxide–semiconductor technology. They are non-suitable as candidates are attractively functionalities of nanoscales are to be realized by utilization of purely energy-efficient electric fields only because the spontaneous and inelastic electromechanical response constitutes their electric polarization. This focus article begins by giving a brief overview of what ferroelectric materials are followed by an introduction to what the challenges entail in relation to the design of thin films as well as a review on the state-of-art in incorporating ferroelectric materials into a variety of electronic products. The ideas concerning chiral states in ferroelectrics, neuromorphic switching, and so on will be introduced in order to be able to see how ferroelectric states can be turned into the demands of the next-generation low-power electronic devices. Lastly, we conclude on the recent progress of the field where a new branch called ferroelectricity at the nanoscale and in 2D system is also being developed [18].

In conclusion, multifunctional materials, advanced manufacturing, electronic and photonic convergence have been named as the antagonists that introduce next-generation electronic. The continuous innovation to overcome the issues of heat dissipation, electromagnetic barrier and the scale-ability of materials is creating a better chance to have a highly integrated and efficient and versatile electronic system.

4.3 CNT-Enhanced Lithium-Ion, Sodium-Ion, and Solid-State Batteries

Due to such extraordinary properties as elevated electro-conductivity, excellent mechanical properties, and large surface areas, carbon nanotubes (CNTs) have taken center-stage in the development of lithium-ion, sodium-ion, and solid-state battery technologies. According to CNTs application in solid-state topologies, that can improve the interfacial contact of the solid electrolyte with the electrode materials, which are usual limiting factors in the performance of solid electrolyte batteries. The rechargeable sodium-ion batteries have attracted much attention due to its advantages like the Availability of sodium reserves and low cost than their commercial counterpart lithium ion batteries.

The flexible solid-state sodium batteries have major potential and usefulness in the polymer matrices owing to their ability to form massive molecular structures with the sodium salts coupled with a low flammable characteristic and high elasticity. Nevertheless, still challenges including inferior ionic conductivity, poor wettability prevent making it possible to retard the performance and application of the battery to some extent. The lithium-ion batteries receive competition mainly from the Sodium-ion batteries, since raw materials are abundantly available and the same geographical location is offered accompanied with cost reduction. The advantages of solid-state sodium ion batteries include low raw material cost, safer property and high energy density, so they are highly attractive to mobile power, electric vehicle and large-scale energy storage fields. The scientific challenges that are considered as posing a serious challenge against solid-state sodium-ion battery and especially solid electrolyte were identified, and the interfacial engineering was also introduced for the improvement of solid-state electrolyte. Industrialization of solid-state Na-ion batteries was demonstrated. The study will be able to expand the knowledge of the scientific problems of the solid-state sodium-ion batteries and provide theoretical insight for the exploitation and utilization of sodium-ion batteries. The paper offers some theoretical advice of much scientific and social significance in the mass commercialization of solid-state sodium-ion batteries. The active materials in carbon nanotubes (CNTs) offer a more homogenous distribution of a battery system and are thus superior in ionic and electrical conductivity of solid-state batteries [19, 20].

With their quite outstanding properties and the appreciated architecture, MXenes another category of transition metal 2D layered group entered the frontlines of energy storage i.e. they are the group of 2D transition metal nitrides and carbides. In this work, the paper talks about the opportunities and barriers of maximizing the energy storage capacities of MXenes in relation to surface functionalization and composite preparation (Fig. 4.1). This work tries to generalize all the ongoing research to define the areas and advancement that would be made in MXene-based supercapacitors [21].

The introduction of carbon nanotubes into the lithium-ion, sodium-ion and solid-state battery enhances the electrical conductivity, structural robustness and ion transport efficiency greatly. Advancements such as these are the foundation of the next generation of

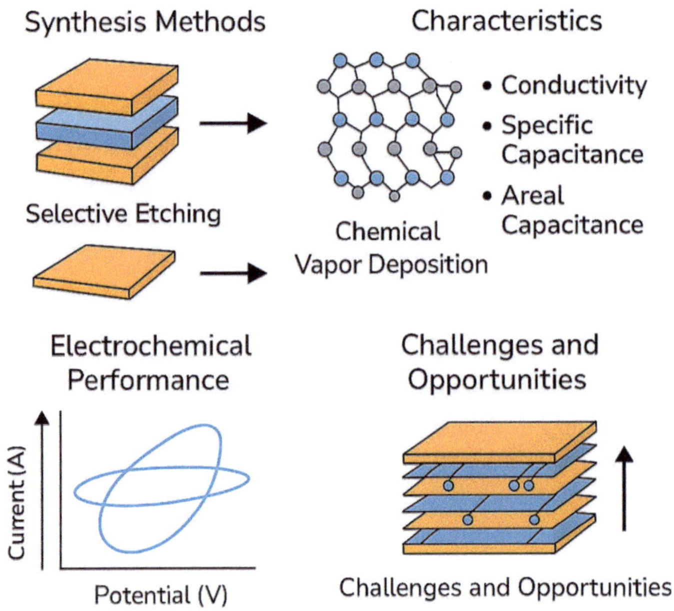

Fig. 4.1 MXenes for supercapacitors

high-performance energy storage systems which would meet the increasing requirements in terms of efficiency and durability of the present-day applications.

4.4 Supercapacitors with High Energy and Power Densities

CNTs are one of the most important components for supercapacitors electrodes material due to their large surface area, high electrical conductance, and good ductility. They are interesting because they can offer supercapacitors increased energy and power densities and are therefore a potential material in a variety of applications such as for portable electronics. The new developments of CNT-based supercapacitor show that the energy density of supercapacitors has been improving greatly. It is hypothesized that the MXene sheets may be used to fill the in-between spaces within CNTs, which leads to a more even filling in the fibers which may enhance strain sensitivity and capacity ability to keep capacitance over long cycle. Such PU@CNT/MXene fibers of great performance and multi-layered hybrid properties are potential candidates of being applied to portable, wearable, and body-integrated sensor systems and energy storage devices. To give an example, the highest energy density of 13.5 mWh cm^{-1} was demonstrated in a study by Hou et al. at a power density of 6.5 mW cm^{-1}, which indicates the prospect of using CNTs to improve the functionality of supercapacitors [22]. The carbon nanotubes (CNTs) have attracted a lot of attention among researchers owing to their outstanding electrical conducting properties, high surface area, and reasonable chemical stability. Here, CNTs have been used to dope with the sulphide based composite $Ag_2S@ZnS$ in different ratios of weight percentages and the composite was prepared through the hydrothermal method. Doping of CNTs positively affected the performance of $Ag_2S@ZnS$ electrode due to their synergetic effect. This device also exhibited better stability, with an initial capacity retention ratio as high as 87% after 5000 cycles. Energy and power densities obtained were 33.02Wh kg1 and 639.98W kg1 respectively with good retention of high amount of coulombic efficiency at 92%. These remarkable performances offer prospects to engineer highly efficient electrode materials of outstanding energy storage devices. Use of CNTs as composite material has also revealed a more effective ion diffusion and the charge storage processes required to attain high energy density [23].

With the current global surge in consumption of energy, and the resultant environmental pollution of energy infrastructure, there has been tremendous amount of pressure on energy resources. Fabrication of more competitive and cheaper binder-free electrodes for the root supercapacitors is of great interest. In this work, two different manners of CNT-E and CNT-NF were blended to enhance carbon reast-sponsive Ni-foam 577 directly made by fully-functional chemical vapor deposition (CVD). Therefore, optimal-CNT electrode electrodes grown so well are used as an electrode material of electrochemical device. For the direct growth of the binder free CNT electrodes research, the paper is significant in explaining the performance of the electrodes in its natural state and the proper storage. Power density is another important factor in the supercapacitor materials, which cannot avoid the interest of CNTs in the former. For instance, the 436.81 F/g specific capacitance of the binder-free-CNT electrode was tested at 1 A/g current density the high-rate state [24], (Fig. 4.2).

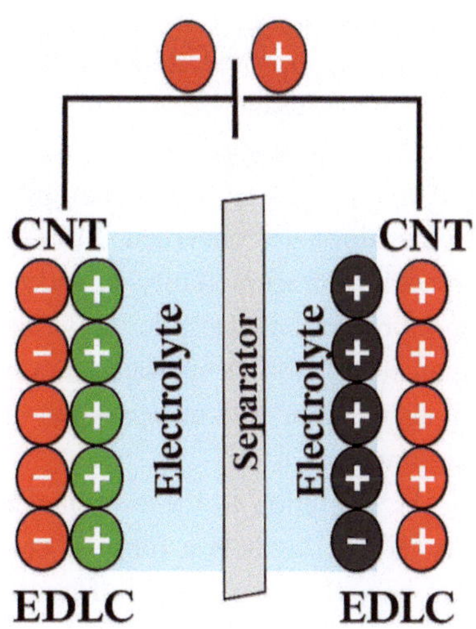

Fig. 4.2 Classification of carbon nanotubes for nanoelectronics

Metal organic frameworks (MOFs) are characterized by advantageous attributes such as large porosity, high capacity of charge storage and active redox sites and they are thus extremely favorable in the ranking of their prospects in the field of energy. Improving the performance of the metal–organic frameworks via modification is essential towards possible widening of the functional application of the same. Nanoparticles (i.e., novel FeNP@MIL-101(Fe)/CNT-based nanocomposites) constructed with material of particular hierarchical micro-mesoporous structures (i.e., Unique Materials provided by Institute Lavoisier (MIL)) have been prepared with appropriate techniques. They were synthesized by facile solvothermal route and their charge storage capabilities were investigated using electrochemical techniques. FeNP@MIL 101(Fe)/CNT composites were first used in supercapacitor electrode as the energy storage device. Moreover, although the study by Mahajan et al. covers the role of carbon nanotubes (CNTs) in providing the overall electrochemical performance of supercapacitors, it does not present accurate information regarding the power densities achieved under CNTs support alone [25]. This retention of high-power density while laying potential for improved energy density by CNTs can give an edge to better supercapacitors.

Finally, the development of carbon nanotube is important in the progress of supercapacitors with elevated energy and power density. The unique individual features foster high electrochemical performance and enable the creation of systems that meet the increasing demand of storage efficiency on energy storage. Existing studies continue putting a focus on the new combinations of carbon nanotubes and other materials that can be used to improve the performance of supercapacitors.

4.5 CNTs as Catalysts and Electrodes in Fuel Cells and Hydrogen Storage

The carbon nanotubes have become useful in the applications of fuel cells and hydrogen storage usage because of their structural configurations and their chemical natures. They have vast surface area and superb mechanical strength as well as excellent electrical conductivity offering them great choices as far as being used as catalytic and storage applications are concerned. Although H_2 has major potential as the clean energy vector, its general embrace will rely on addressing the major challenges in load efficiency and safety in its operations. Despite the value of hydrogen as a clean energy carrier, many existing methods of storing hydrogen in a form that is usable for transportation or heating involve liquefaction or cryo-compression of hydrogen, which are typically not economical, do not scale well.

As a solution to these obstacles, researchers are already exploiting nanotube materials (NTMs) as it is a form of crystalline composition with high unique properties that can be used in the storage of hydrogen. Structural flexibility—The non-thermal materials can be critically prepared and closely configured to allow increased absorption of hydrogen. These materials show significant porosity having large capacities in accommodating hydrogen molecules. The large surface area assists in the higher hydrogen adsorption capability of NTMs. Weaker van der Waals forces-based technologies adsorb hydrogen by taking advantage of weak forces and then liberate hydrogen either by heating or reducing pressure. Efforts are on the way to enhance the surface area and hydrogen absorption of NTMs with high density resulting in storage of hydrogen through procedures like the spill-over of hydrogen. The ability to bind and hold onto the molecules of hydrogen is commonly analyzed in transition metals within non-conventional materials. This addition can reduce the gravimetric density of the material; a very important practical consideration. To sum it up, research aimed at studying NTMs and their ability to store hydrogen based on density functional theory is being conducted. The reason behind this study is in order to enhance hydrogen storage, especially by doping transition metal nanostructured materials. Even through these metals improve hydrogen adsorption, consideration should be paid to compromises in gravimetric density. The only adsorption mechanisms of hydrogen on CNTs are physisorption that hydrogen molecules are physiosorbed on the CNTs surface due to weak van der Waals forces, and they could be released readily through heating or decompressing etc. [26].

Thanks to their high hydrogen storage ability and relatively low cost as hydrogen storage materials, magnesium hydride (MgH_2) has been long considered as a candidate for hydrogen storage. Nevertheless, it is impractical to use it because of low thermal stability and a slow rate of dehydrogenation. In this regard, a bifunctional catalyst: carbon nanotubes (CNTs) and palladium (Pd) was prepared through a hydrothermal and calcination procedure, which significantly improved the hydrogen adsorption–desorption kinetics of Mg/MgH_2 system. The addition of such a CNT Pd catalyst lowered activation energy required to absorb hydrogen to 42.73 kJ/mol, which was obtained during implementation of the Avrami-Erofeev model and the application of the Arrhenius equation. What is more, the Avrami-Kolmogorov theory and Arrhenius analysis were used to determine the dehydrogenation activation energy amounting to 73.98 kJ/mol. These positive aspects confirm the efficiency of the catalyst in accelerating and making energy-efficient the release and intake of hydrogen in the Mg-made system. The dynamic optimization persists following 10 cycles. The ground-breaking finding has supported future applications of MgH_2 as a storage medium of hydrogen energy, presenting increased growth opportunities. Besides, chemical mixing of metal dopants such as palladium or nickel with carbon nanotubes (CNTs) has been demonstrated with improved hydrogen adsorption capacity and enhancement of hydrogen uptake/release rate of hydrogen absorption and desorption [27].

They synthesized Ni-Cu/MSN (Ni 50%, Cu 5%), Ni–Fe/MSN (Ni 50%, Fe 5%), and Ni/MSN (Ni 50%), bimetallic catalysts using the impregnation technique and used each of them to investigate and compare their performances. In the framework of the methane decomposition, the long-term activity and the catalytic performance of them were studied. In addition to that, Fe and Cu promoter's role was investigated by examining their effect in a temperatures range between 700 and 900 degrees Celsius. The results shown that addition of Cu and Fe to the catalysts had significant positive contribution to the methane-to-hydrogen yield compared to unpromoted catalysts. This is because the catalysts have been deposited with Ni-Cu and Ni–Fe bimetallic alloy, which have better stability. Hydrogen production also increases with the reactant temperature increase. The results suggested that the introduction of iron (rather than that of copper) in the Ni/Fe alloy resulted in enhanced H_2 production, and in the longer life of the nickel phase at all the reaction temperatures. The analysis of this wasted catalyst also supported the morphological features of bamboo-like internal, a well crystalline and certain graphitized of the produced carbon multi-walled carbon nanotubes. To introduce the cleavage of methane to hydrogen and carbon, doping with transition metals would increase the catalytic efficiency of CNTs (carbon nanotubes) to provide a more environmentally friendly method for hydrogen production [28].

Due to the complex structures and impressive physicochemical properties of CNTs, the carbon nano-tubes (CNTs), appellation which represents an allotrope of carbon, have been introduced in recent years in various fields including energy source, chemical production and environmental protection. Functionalized carbon nanotubes are particularly desirable in the fabrication of composites of different components which are particularly desirable in the extraction of CO2 from the atmosphere and from within the body, for the reason that they have a complicated pore structure and reactive sides. This paper presents a detailed analysis of carbon nanotube synthesis techniques, including the major nootropic chemical vapour deposition, arc discharge, and even laser-ablation, and their influence on carbon-nanotube precursor and its structure and properties. Herein, applications of CNT in fuel cell, photosynthesis, ammonia production, dry methane reformation, fishertropsch production and supercapacitor have been highlighted. However, application of carbon nanotubes has been limited because their manufacturing process is complex and expensive, making it difficult to commercially mass-produce them. The challenges associated with the production of carbon nanotube composites and the opportunities of their development are described in the end of this paper. Carbon nanotubes (CNTs) have also been explored as catalysts in fuel cells as this would result in an increased overall efficiency of the electrochemical reactions involved in fuel cells [29].

To conclude, carbon nanotubes ful-fill two functions as catalysts and material to store hydrogen, which makes them necessary when it comes to developing effective fuel cell technologies as well as developing sustainable hydrogen storage systems. Their peculiarities allow significant adsorption of hydrogen and enhance catalytic processes and make them significant components in the transition to a hydrogen economy.

4.6 Challenges and Ethical Considerations

As much as CNTs have distinct physical benefits that make its application widespread, there are safety, environmental and ethical responsibility aspects that remain unresolved therefore presenting the case in which CNTs need thorough examination and regulation. One of the greatest issues concerning the CNTs is the complex and consequently costly production schemes. Despite being rather interesting in terms of their possible applications, the issues related to the preparation of CNT composites hinder the production of such materials under an industrial scale. This limitation does not only discredit economic viability of CNTs, but raises concerns regarding the ecological consequences of their production techniques. This is because the manufacturing techniques commonly involve the utilization of toxic substances and resources harnessed consumptively, which may lead to environmental degradation by failure to manage them [29, 30]. According to the research, the artificially produced nanomaterials, including carbon nanotubes (CNTs), can raise concerns about human health either through occupational exposures when they are synthesized and handled [31, 32]. These concerns are aggravated by the lack of incisive evidence regarding exposure concentrations and their effects throughout the life cycle of CNTs and thus a long-term risk assessment and control measures to protect not only workers, but also consumers are required. Other than health hazards, the environmental impacts of CNTs may be severe. The accumulation of carbon nanotubes in the ecosystems can lead to the unforeseen consequences, including the toxicity of aquatic life and potential bioaccumulation along food chains [33, 34]. Disposing of products that include carbon nanotubes is a challenge, as they did not apply normal waste management procedures, which fail to defend the special properties of nanomaterials, and thus cause pollution to the environment. The toxicity and biocompatibility of CNTs in biomedical use are also some of the ethical concerns. The surrounding environment is increasingly being exposed to nanoparticles as a result of many nanotechnologies, which can cause potentially harmful effects to human health and environmental safety. Therefore, it is pressing to take these harmful particles off the air immediately. In the present study, the potential of Trametes versicolor in the bio-remediation of carboxylate functionalized Multi-Walled Carbon Nanotubes (MWCNTs) was studied. It was found that liveT. versicolor biomass displayed better NPs removal capacity than dead one under both 30 °C and pH7. Increasing nanotube concentration, 5 to 20 mg. mL^{-1} Biosorption capacity decreased adsorption efficiency from 100 to $28.55 \pm 1.7\%$. Moreover, the study also indicated that there is increase in adsorption capacity and biomass production (which is enhanced biosorption ability for concentrations > 5 mg) accompanying an extension of initial fungal growth time. ml − 1. Carbon nanotubes are a new type of carbon nanomaterials that are widely used in the biotechnology and biomedicine research, especially in neurological disorder. Finally, the challenges of carbon nanotube clinical testing and how these might be addressed are discussed. Despite the potentialities of CNTs in drug delivery and oncology treatment, the fact that they interact with biological systems raises the alarm on their safety and efficacy [35–37].

To summarize, despite the fact that carbon nanotubes offer the great potential to a variety of industries, the problems and ethical concerns that the technology raises cannot indeed be overlooked. The details of the manufacturing, potential health risk, environmental impact, and biocompatibility are the must-to-be addressed factors towards the responsible development of CNT technology. Further researches should focus on the development of safer synthesis methods, comprehensive risk assessments and robust regulatory framework that will guide the practical application of CNTs in industries and the medical field into the future.

4.7 Ultra-Lightweight and High-Strength Materials with Multifunctional Properties

Minor component Carbon nanotubes (CNTs) have proved to be an important additive in the synthesis of ultra-light, high strength multifunctional composites. The unique structural properties, including a high aspect ratio, exceptional mechanical properties and light weight composition, which makes them the best-suited in most technological services, particularly aerospace, automotive and electronics companies. Among the most important properties of CNTs, one may distinguish its unbelievable mechanical properties. Having a tensile strength over 100 GPa, Young moduli above 1 TPa and a strain to fracture of more than 2%, these materials outperformed in a significant margin the traditional materials [38]. Such amazing ratio of strength to weight allows the creation of the composite, which is not only light but also strong enough to resist great amount of mechanical load. Studies have shown that incorporation of carbon nanotube into a polymer composite could enhance the mechanical strength of these materials and make them suitable in high-performance applications [39, 40]. Addition of carbon nanotube to various matrices has led to materials with high structural integrity and effective electromagnetic interference shielding, making those suitable in the aerospace industry or the military. Research proves that carbon nanotubes can significantly improve the heat transfer conduction of polymers, which is crucial to be applied in heat-sensitive conditions [41–44]. The property is particularly beneficial to the automobile industry where light materials of higher thermal stability are necessary to improve fuel consumption and performance. This adaptability enables the production of the composites that suit the needs of specific industries, all the way to construction and electronics [45, 46]. Overall, it appears that by using carbon nanotubes as composite material an ultra-lightweight and high-strength multi-functional material can also be produced. It is carbon nanotubes which have impressive kinetical properties and their characteristics can be tailored through functionalization and are therefore central in the creation of next-generation materials in a wide spectrum of applications.

4.8 CNT-Based Water Purification and Desalination Membranes

Due to the unique crystalline structure and chemical feature, carbon nanotubes (CNTs) are being identified as a new class of advanced nanomaterials promising for the application in filtering and desalination. The high surface area, mechanical strength and ability to generate nanopores render them a potential candidate for improving the performance of such membranes for water treatment. Due to the unique crystalline structure and chemical feature, carbon nanotubes (CNTs) are being identified as a new class of advanced nanomaterials promising for the application in filtering and desalination. The high surface area, mechanical strength and ability to generate nanopores render them a potential candidate for improving the performance of such membranes for water treatment. The hydrophobic nature of CNTs increases its performances in rejecting the organic contaminants by developing some barrier that reduces the entry of water molecules into the membrane structure. This underscores the necessity of optimizing the process of the membrane structure in order to find an optimum condition of a trade-off between porosity and permeability that is vital to effective water treatment. The latest molecular resimulations shed light on the hydrodynamics of the process of water transport along carbon nanotubes (CNTs) and confirmed that the diameter and the hydrophobicity of the tubes also significantly influence the overall behavior of the water flow [47–49].

This functionalization enhances the surface adsorption and the interaction of the CNTs with the pollutants, hence, greater efficiencies of removing them. To sum up, CNT based membranes mark a major breakthrough in the water purification and desalination field. These unique properties including the higher surface area, mechanical properties, and hydrophobicity can be modified make them suitable in water treatment in various applications. The further exploration of the CNT membrane enhancement process through its functionalization and structural modification is likely to offer even more effective ways of reducing the global problems with water deprivation.

4.9 Air Filtration and Pollutant Removal Applications

The structural, electrical performance and heat dissipation capabilities have made carbon nanotubes (CNTs) an all-round and productive air filtration feed and pollution removal constituent. Their large -scale interface area, high mechanical strength and allowance to functionalize them are particularly effective in the remediation of the environment e.g. extraction of various contaminating sources in air and water. The foremost strength of CNTs in air filtration is that it contains sizeable adsorption landing of various pollutants. Functionalization of carbon nanotubes may immensely improve how carbon nanotubes will interact with certain pollutants. GO contains extensive numbers of oxygen-bearing functional groups (e.g. hydroxyl, carboxyl, and epoxy), improving its hydrophilicity and the ability to physically adsorb cationic pollutants. Some mechanisms of these nanomaterials were reviewed and that include; Physical adsorption due to van der Waals attraction energy and pi-pi interatomic bonding, and chemical adsorption in which nanomaterials adsorb through covalent or ionic bonding with the pollutants. This report presents the significant challenges of nanoparticle stability, recovery, cost of production, environmental and health impact of applying these few nanoparticles in practice. Some of the key issues such as clumping of nanoparticles1, re-usability of particles etc., safety of exhausted material are addressed and potential findings include surfaces coated with nanoparticles for use as anti-clumping agents for enhancing particle dispersion transport within the water body and use of magnetic nanoparticles to make the collection of nanoparticles easier. The results support the idea of the tremendous opportunities in GO and CNTs in developing environmental remediation technology, which can open a road toward the development of more effective and cleaner solutions to water treatment. With the further development of the field, the prospects of the interdisciplinary cooperation and further investigation of the issue guarantee significant improvements in the field of water purification and environmental protection [50]. Additionally, carbon nanotubes can also be impregnated into other materials, e.g. graphene oxide, to create multiphase materials, and this has a better adsorption potential, and, therefore, an overall performance increase in the environmental context. In addition to that, CNTs have been also used in the photocatalytic system where it could raise the degradation of organic pollutants under the light influence. Photocatalytic composites made of CNTs with a functionalization have proven to have a higher degradation rate on dye and other organic pollutants in the wastewater [51, 52].

Plasticity of the CNTs could embrace the use of the CNTs in hybrid materials where they could be mixed with additional nanomaterials to produce multifunctional mixtures to enhance the absorption of more pollutants. In this connection, MXene/CNT composites were widely advanced because of high environmental stability, good electrical performance because of their great mechanical performance they exhibited strong potentials in the development of innovative systems both in the diagnosis, therapy, environmental and electronic sector. The MXenes are unique in structure, a large specific surface area, the easy graft ability of functionalization, and the high electrical conductivity. They are deployed as composite material together with carbon nanotubes (CNTs) that as a one-dimensional conductive nanomaterial have not only admirable thermal and electric conductivities but also better mechanical properties. The vast majority of the research is nevertheless related to their electrical, EMI shielding, catalytic, and sensing properties; there is, therefore, a strong need to investigate their biological and diagnostic/therapeutic features. In this connection, MXene/CNT composites were widely advanced because of high environmental stability, good electrical performance because of their great mechanical performance they exhibited strong potentials in the development of innovative systems both in the diagnosis, therapy, environmental and electronic sector. The MXenes are unique in structure, a large specific surface area, the easy graft ability of functionalization, and the high electrical conductivity. The vast majority of the research is nevertheless related to their electrical, EMI shielding, catalytic, and sensing properties; there is, therefore, a strong need to investigate their biological and diagnostic/therapeutic features [53, 54].

These coupled systems can be customized to suit certain contaminants hence making the remediation operations more effective at large scale. In conclusion, carbon nanotube properties (and its capability to become functionalized) is what makes it possible to use the nanotubes to filter and trap items in the air. The fact that they possess high surface area and that they can be photo catalytically active and be employed in the complex filtering systems make yet another significant material in the continued mission to reduce the pollution issue.

4.10 Green Synthesis Routes to Reduce Toxicity and Environmental Impact

Green synthesis of carbon nanotubes is attracting attention because of increased attention towards green and environmentally friendly industrial production. Traditional methods normally involve the usage of risky catalyst and power-intensive processes, which have cost the environment and cause a health risk. By contrast green synthesis processes exploit renewable resources, e.g. biomass, and aim to minimize toxicity and impact to the environment. One of such ways is the utilization of biomass into manufacture of carbon nanotubes. Duraia et al. demonstrated efficient green synthesis of carbon nanotube based on yellow maize by a one-step pyrolysis process, which aims at reducing environmental footprint and utilization of agricultural waste, therefore, establishing a circular economy. It's important to remark that even though CNTs can be obtained by the pyrolysis of biomass, the quality of the resulting CNTs may be an issue, not to mention that the process can demand the inclusion of nanoparticulate catalysts that may contaminate the final product. Qasim et al. have highlighted the advantage associated with using biomass to make carbon nanoparticles by emphasizing the importance of low-toxicity production processes that meet the threshold of sustainability [6, 55]. The studies point out the potential of CNTs derived by use of biomass as a viable alternative to the conventional methods of synthesis.

The other innovative technique applies the utilization of lignin which is a by-product of paper and biofuel industries as a raw material to produce carbon nanotubes. The method produces higher-quality carbon nanotubes and leads to the increased sustainability of the energy storage sector because of the reuse of lignin. The use of lignin in manufacturing carbon nanotube is used to describe the utilization of wastes to make nanomaterials useful, hence reduce environmental impact. CNT structural tuning and chemical tailoring also fall into the practice of green chemistry. Deposition of functional molecules is extensively used to modulate the optical and electrical properties of the assembled CNTs in an environmentally friendly way without the use of toxic agents by surface modification in terms of aryl diazonium chemistry [56, 57]. Such a strategy enhances the performance of CNTs and meets the requirements of the environmentally friendly standards, which demonstrates the flexibility of the green synthesis approaches.

Other than the biomass and lignin, a number of natural sources and methods are under research to develop carbon nanotubes (CNT). However, the initial passages have brought no specific references to support this claim. Finally, to curb the effects of toxicity and environmental impact, the promotion of environmental-friendly synthesis procedures in carbon nanotubes is highly required. Scientists are continuing developing the environmentally friendly synthesis of nanomaterials; they also rely on renewable raw materials and ecofriendly chemical processes to synthesize nanomaterials. These advances do not only lead to environmental protection but also boost the circular economy as those products are made out of garbage.

4.11 Conclusion Remarks

The sphere of carbon nanotubes (CNT) examination has experienced an amazing move, especially concerning the procedure of synthesis and purification. Such methodological improvements have not just made the production of CNTs more reliable and consistent, but also made them much more applicable in advanced forms of technology. The incorporation of CNTs in the next generation electronics devices would be the ultimate form of a leap, owing to its ability to facilitate enhancements in performance, miniaturization, and multifunctionality, as the conventional materials are unable to accomplish. CNTs have been applied in the energy storage field as the key materials in lithium-ion, sodium-ion, and solid-state batteries, where its extraordinary electrical conductivity and mechanical strength play improved capacity, cycling stability and safety. In addition to energy storage, CNTs can also be a useful catalyst and electrode in fuel cells and in hydrogen storage systems thus demonstrating again their utility and promise in terms of green energy applications. However, a considerable proportion of difficulties is coming along with the wide usage of CNT based technologies. Scalability of manufacturing that maintains uniformity of the materials is also a major bottleneck. In addition, environmental ethics and human health outcomes will also require stringent evaluation models in facilitating responsible innovation. The unique key intrinsic characteristics of CNTs, its ultra-light and at the same time very strong materials, provide unprecedented possibilities to design multifunctional materials that satisfy the demanding functions of the current applications. The new studies of CNT-based membranes in water treatment and desalination point to the potential of the material to meet the urgent needs of the global environment. On the same note, the use of CNTs in air filtration and the elimination of pollutants also enhances the applications of CNTs in the field of environmental remediation which is in tandem with the material science advancements towards the necessities required in sustainability. Notably, the green synthesis routes formation obviates environmental hazards, decreases toxicity and leads to such environmentally friendly production manners. To conclude, although carbon nanotubes present a technological breakthrough on the field of nanotechnology-based innovation, their potential is best utilized upon a synergetic field positioning, between the synthesis, application, and ethical control domains. There should be continuous cross-cutting research and development in the fields of materials science, environmental engineering, and policy to use laboratory-based success stories to scale and make them safe and effective technologies. This kind of integration will not only stimulate the technological development but it will also lead to substantial contributions to the achievements of sustainable development, in which CNTs become the key materials in shifting the world toward a more efficient and ecofriendly technological environment.

References

1. Sumana, V., Sudhakar, Y., Nagaraja, G., Savitha, M., & Puthiyillam, P. (2024). Exploring the potential of poly (caprolactone) and guar gum biodegradable blend film: an investigation for supercapacitor. Materials Research Express, 11(5), 055303. https://doi.org/10.1088/2053-1591/ad4199
2. Vishwanath, M. (2023). Secondary electron yield decreases of laser drilled microporous gold surfaces: empirical modelling and monte carlo simulation. https://doi.org/10.31219/osf.io/egza9
3. Milenov, T., Trifonov, D., Kalchevski, D., Kolev, S., Avramova, I., Russev, S., Genkov, K., Avdeev, G,, Dimov, D., Karaivanova, D.M. & Valcheva, E. (2023). Study of the chemical vapor deposition of nano-sized carbon phases on {001} silicon. Materials, 16(22), 7190. https://doi.org/10.3390/ma16227190
4. Ushakov, I. and Oshorov, A. (2022). Viscosity of micro destruction of multilayer composite and method of its revealing. Materials Science Forum, 1052, 110–115
5. Liu, F., Yuan, M., He, Y., Zhai, G., Xiang, H., Wu, Q. & Zhu, M. (2024). Insights into the nucleation and structure of lignin-based carbon nanotubes synthesized using iron via floating catalyst chemical vapor deposition. Acs Applied Nano Materials, 7(7), 7184–7194. https://doi.org/10.1021/acsanm.3c06235
6. Duraia, E., Opoku, M., & Beall, G. (2024). Efficient eco-friendly synthesis of carbon nanotubes over graphite nanosheets from yellow corn: a one-step green approach. Scientific Reports, 14(1). https://doi.org/10.1038/s41598-024-65893-6
7. Báti, G., Laxmi, S., & Stuparu, M. (2023). Mechanochemical synthesis of corannulene: scalable and efficient preparation of a curved polycyclic aromatic hydrocarbon under ball milling conditions. Chemsuschem, 16(20). https://doi.org/10.1002/cssc.202301087
8. Martinčić, M. and Tobías, G. (2024). Uv-vis quantification of the iron content in iteratively steam and hcl purified single-walled carbon nanotubes. Plos One, 19(5), e0303359. https://doi.org/10.1371/journal.pone.0303359
9. Wu, B., Zhu, H., Yang, Y., Zhu, K., Liu, R., & Li, Y. (2023). Hierarchically flexible polyamide composite films for thermal management and electromagnetic shielding. Acs Applied Engineering Materials, 1(1), 568–576. https://doi.org/10.1021/acsaenm.2c00138
10. Du, K., Meng, Z., Xi, Y., Liu, N., Zhang, J., Xu, S., Shi, Z., Zhang, H., Wang, Sh., Feng, H., Hao, W., Pan, H., Zhang, Sh. & Du, Y. (2024). Controllable modulation of the electronic properties of a two-dimensional ambipolar semiconductor by interface ferroelectric polarization. ACS Applied Materials & Interfaces, 16(3), 4181–4188. https://doi.org/10.1021/acsami.3c15191

11. Cai, Z., Liu, L., & Zhou, P. (2024). The development of transfer technologies for advanced 2d circuits integration. Information & Functional Materials, 1(3), 304–322. https://doi.org/10.1002/ifm2.28
12. Guo, T., Zhou, D., Gao, M., Deng, S., Jafarpour, M., Avaro, J., Neels, A., Hack, E., Wang, J., Heier, J. & Zhang, C. (2023). Large-area smooth conductive films enabled by scalable slot-die coating of TI3C2T_X mxene aqueous inks. Advanced Functional Materials, 33(15). https://doi.org/10.1002/adfm.202213183
13. Guo, X., Hu, X., Zhang, S., Yang, J., Chen, C., Zhang, J., Qu, H., Zhang, Sh. & Zhou, W. (2023). High-performance and low-power p-channel transistors based on monolayer Be_2C. ACS Applied Materials & Interfaces, 15(46), 53644–53650. https://doi.org/10.1021/acsami.3c09470
14. Kim, M., Park, J., Cho, C., & Ko, S. (2023). Liquid metal based stretchable room temperature soldering sticker patch for stretchable electronics integration. Advanced Functional Materials, 33(36). https://doi.org/10.1002/adfm.202303286
15. Yang, J., Zhang, Z., Zhou, P., Zhang, Y., Liu, Y., Xu, Y., Gu, Y., Qin, Sh., Haick, H. & Wang, Y. (2023). Toward a new generation of permeable skin electronics. Nanoscale, 15(7), 3051–3078. https://doi.org/10.1039/d2nr06236d
16. Guchait, A., Pramanik, S., Goswami, D., Chattopadhyay, S., & Mondal, T. (2024). Elastomeric ionic hydrogel-based flexible moisture-electric generator for next-generation wearable electronics. ACS Applied Materials & Interfaces. https://doi.org/10.1021/acsami.4c11907
17. Pérez-Díaz, O., Estrada-Wiese, D., Aceves-Mijares, M., & González-Fernández, A. (2023). Functionalization of a fully integrated electrophotonic silicon circuit for biotin sensing. Biosensors, 13(3), 399. https://doi.org/10.3390/bios13030399
18. Müller, M., Efe, İ., Sarott, M., Gradauskaite, E., & Trassin, M. (2023). Ferroelectric thin films for oxide electronics. Acs Applied Electronic Materials, 5(3), 1314–1334. https://doi.org/10.1021/acsaelm.2c01755
19. Lin, Z. and Wang, Z. (2023). Application of solid polymer electrolytes for solid-state sodium batteries. Matec Web of Conferences, 386, 03019. https://doi.org/10.1051/matecconf/202338603019
20. Si, H., Ma, J., Xiao, X., Wang, Q., Geng, S., & Fu, L. (2025). Solid-state sodium-ion batteries: theories, challenges and perspectives. Chemistry—A European Journal, 31(5). https://doi.org/10.1002/chem.202403247
21. Singh, A., Lalotra, N., Shah, S. S., Rather, M. D., López-Maldonado, E. A., Pathania, K., Abaszade, R. G., Stetsenko, M., & Arya, S. (2025). A comprehensive review on the synthesis and properties of MXenes for supercapacitor applications. *ACS Applied Energy Materials*, 8(8), 4884–4914. https://doi.org/10.1021/acsaem.4c03194
22. Hou, K., Xin, S., Yuan, J., Zhang, D., Bai, Y., Xu, J., Guo, F., Zhang, Y., Cao, A. & Shang, Y. (2023). Stretchable Pu@CNT/MXene fiber fabricated by external tension for high performance strain sensor and supercapacitor. Advanced Materials Technologies, 8(18). https://doi.org/10.1002/admt.202300479
23. Rehman, A., Afzal, A., Iqbal, M., Ali, M., Wabaidur, S., Al-Ammar, E., Imran, M., Mumtaz, S., Choi, E.H. & Muzaffar, N. (2023). Synergetic and anomalous effect of CNTs in the sulphide-based binary composite for an extraordinary and asymmetric supercapacitor device. The Canadian Journal of Chemical Engineering, 102(1), 226–241. https://doi.org/10.1002/cjce.25048
24. Isacfranklin, M., Yuvakkumar, R., Ravi, G., & Velauthapillai, D. (2023). Direct growth of binder-free cnts on a nickel foam substrate for highly efficient symmetric supercapacitors. ACS Omega, 8(13), 11700–11708. https://doi.org/10.1021/acsomega.2c04998

25. Mahajan, H., Shah, A., Kim, S., & Cho, S. (2024). Fenp@mil-101(fe)-based carbon nanotube composite for energy storage applications. ACS Omega, 9(23), 24546–24557. https://doi.org/10.1021/acsomega.4c00590
26. Villagracia, A. (2024). A comprehensive bibliometric review and analysis on the evolution of nanotube-based hydrogen storage materials via DFT simulations. IOP Conference Series Earth and Environmental Science, 1372(1), 012098. https://doi.org/10.1088/1755-1315/1372/1/012098
27. Duan, C., Wu, J., Liang, P., Wang, H., Qu, T., Fan, Y., Su, Y., Su, Zh., Hu, L. & Wu, Y. (2024). Cnts-pd as efficient bidirectional catalyst for improving hydrogen absorption/desorption kinetics of Mg/MgH_2. Energy & Fuels, 38(10), 8979–8991. https://doi.org/10.1021/acs.energyfuels.4c01138
28. Hasnan, N., Pudukudy, M., Yaakob, Z., Kamarudin, N., Lim, K., & Timmiati, S. (2023). Promoting effects of copper and iron on ni/msn catalysts for methane decomposition. Catalysts, 13(7), 1067. https://doi.org/10.3390/catal13071067
29. Tao, Z., Zhao, Y., Wang, Y., & Zhang, G. (2024). Recent advances in carbon nanotube technology: bridging the gap from fundamental science to wide applications. C – Journal of Carbon Research, 10(3), 69. https://doi.org/10.3390/c10030069
30. Jia, Q. (2023). Synthesis and modification of carbon nanotubes. Applied and Computational Engineering, 26(1), 115–119. https://doi.org/10.54254/2755-2721/26/20230810
31. Fito, C., Colmenar, I., Andreu-Sánchez, Ó., Vela, V., & Domat, M. (2023). Exposure assessment and risk characterization of carbon-based nanomaterials at different production scales. Sustainability, 15(16), 12544. https://doi.org/10.3390/su151612544
32. Berthing, T., Lard, M., Danielsen, P., Abariūtė, L., Barfod, K., Adolfsson, K., Knudsen, K.B., Wolff, H., Prinz, Ch. N. & Vogel, U. (2023). Pulmonary toxicity and translocation of gallium phosphide nanowires to secondary organs following pulmonary exposure in mice. Journal of Nanobiotechnology, 21(1). https://doi.org/10.1186/s12951-023-02049-0
33. Gendron, D. and Bubak, G. (2023). Carbon nanotubes and graphene materials as xenobiotics in living systems: is there a consensus on their safety?. Journal of Xenobiotics, 13(4), 740–760. https://doi.org/10.3390/jox13040047
34. Aldosari, H. (2023). Review of carbon nanotube toxicity and evaluation of possible implications to occupational and environmental health. Nano Hybrids and Composites, 40, 35–49. https://doi.org/10.4028/p-4gnl3o
35. Asefi, S. and Moghimi, H. (2023). Removal of carboxylated multi-walled carbon nanotubes (MWCNT-COOH) from the environment by trametes versicolor: a simple, cost-effective, and eco-friendly method. Scientific Reports, 13(1). https://doi.org/10.1038/s41598-023-43517-9
36. Gao, S., Xu, B., Sun, J., & Zhang, Z. (2024). Nanotechnological advances in cancer: therapy a comprehensive review of carbon nanotube applications. Frontiers in Bioengineering and Biotechnology, 12. https://doi.org/10.3389/fbioe.2024.1351787
37. Wilson, D., Ahlawat, J., & Narayan, M. (2024). Carbon nanotubes as neuroprotective agents. Explor Neuroprot Ther, 72–81. https://doi.org/10.37349/ent.2024.00071
38. Cho, Y., Lee, J., Jung, Y., Park, J., Park, J., Kim, S., Yang, S. J. & Park, C. (2024). Supertoughness carbon nanotube yarns by bio-inspired nano-coiling engineering. Advanced Science, 11(25). https://doi.org/10.1002/advs.202400460
39. Tahhan, A., Alkhedher, M., Ramadan, M., & Al-Adwan, I. (2023). Investigating the effect of structural modifications on mechanical properties of carbon nanotubes under tensile loading using molecular dynamics simulations. Key Engineering Materials, 962, 45–52. https://doi.org/10.4028/p-kmfq7j

40. Hong, S., Lee, O., Lee, H., & Yoon, Y. (2023). Evaluation of flexural strength of high-strength lightweight cement composites containing carbon nanotubes. Korean Society of Hazard Mitigation, 23(4), 1–9. https://doi.org/10.9798/kosham.2023.23.4.1
41. Atay, H. and Bilgiç, N. (2024). Radar absorbing properties of different size carbon nanotube reinforced polymer composites. Frontiers in Materials, 11. https://doi.org/10.3389/fmats.2024.1380472
42. Kim, S., Lee, S., Zhang, Y., Park, S., & Gu, J. (2023). Carbon-based radar absorbing materials toward stealth technologies. Advanced Science, 10(32). https://doi.org/10.1002/advs.202303104
43. Najmi, L. and Hu, Z. (2024). Effects of topological parameters on thermal properties of carbon nanotubes via molecular dynamics simulation. Journal of Composites Science, 8(1), 37. https://doi.org/10.3390/jcs8010037
44. Alshameri, M., Aziz, A., Arshad, A., & Sadeq, S. (2025). The effect of carbon nanotube-reinforced polymer fins in compact heat exchanger. Journal of Physics Conference Series, 2933(1), 012033. https://doi.org/10.1088/1742-6596/2933/1/012033
45. Yang, S. (2023). Properties, applications, and prospects of carbon nanotubes in the construction industry. Architecture Structures and Construction, 3(3), 289–298. https://doi.org/10.1007/s44150-023-00090-z
46. Ran, Z. (2023). Application of carbon nanotube composite materials in aviation. Applied and Computational Engineering, 23(1), 177–182. https://doi.org/10.54254/2755-2721/23/20230649
47. Mendoue, D., Benelfellah, A., Boumbimba, R., Royaud, I., Ponçot, M., Wary, M., & Noirel, C. (2024). Study of the thermomechanical behavior of composite based on elium acrylic resin, carbon nanotubes, and flax fibers: experimental approach. Polymer Composites, 45(13), 11785–11803. https://doi.org/10.1002/pc.28597
48. Sahu, P. and Ali, S. (2023). Uniqueness of nanoscale confinement for fast water transport: effect of nanotube diameter and hydrophobicity. The Journal of Physical Chemistry B, 128(1), 222–243. https://doi.org/10.1021/acs.jpcb.3c05979
49. Sahu, P. and Ali, S. (2024). Studies on continuum breakdown and water transport behavior of nanotubes for water purification: impact of temperature, nanotube material, and diameter. ACS Applied Nano Materials, 7(8), 9228–9243. https://doi.org/10.1021/acsanm.4c00729
50. Younas, A., Rashad, M., Rafaqat, A., Maqbool, H., Gul, K., Ahmad, M., Shahbaz, J., Kanwal, K. & Umair, M. (2024). Adsorption of organic compounds and hazardous microorganisms from sewage and surfactant-containing wastewater using carbon-based nanomaterials: a focus on graphene oxide and carbon nanotube. Biological and Clinical Sciences Research Journal, 2024(1), 1052. https://doi.org/10.54112/bcsrj.v2024i1.1052
51. Bhayana, P., Dandekar, T., Bala, R., Madaan, R., & Mahajan, P. (2024). Role of carbon nanotubes in remediation of pharmaceutical wastewater. Remediation Journal, 34(2). https://doi.org/10.1002/rem.21776
52. Ghourichay, S., Agbolaghi, S., Corpino, R., & Ricci, P. (2024). Carbon nanotube–phenyl modified g-c3n4: a visible light driven efficient charge transfer system for photocatalytic degradation of rhodamine b. Molecules, 29(22), 5439. https://doi.org/10.3390/molecules29225439
53. Mohajer, F., Ziarani, G., Badiei, A., Iravani, S., & Varma, R. (2023). MXene-carbon nanotube composites: properties and applications. Nanomaterials, 13(2), 345. https://doi.org/10.3390/nano13020345
54. Hughes, K., Iyer, K., Bird, R., Ivanov, J., Banerjee, S., Georges, G. & Zhou, Q. (2024). Review of carbon nanotube research and development: materials and emerging applications. ACS Applied Nano Materials, 7(16), 18695–18713. https://doi.org/10.1021/acsanm.4c02721

55. Qasim, M., Clarkson, A., & Hinkley, S. (2023). Green synthesis of carbon nanoparticles (CNPS) from biomass for biomedical applications. International Journal of Molecular Sciences, 24(2), 1023. https://doi.org/10.3390/ijms24021023
56. Liu, C., Ma, Y., Xie, Y., Zou, J., Wu, H., Peng, S., Wei, Q., He, D., Zhang, X., Li, B.-W. & Nan, C. (2023). Enhanced electromagnetic shielding and thermal management properties in mxene/aramid nanofiber films fabricated by intermittent filtration. ACS Applied Materials & Interfaces, 15(3), 4516–4526. https://doi.org/10.1021/acsami.2c20101
57. Heppe, B., Dzombic, N., Keil, J., Sun, X., & Ao, G. (2023). Solvent isotope effects on the creation of fluorescent quantum defects in carbon nanotubes by aryl diazonium chemistry. Journal of the American Chemical Society, 145(47), 25621–25631. https://doi.org/10.1021/jacs.3c07341

Index

R. G. O. Abaszade and A. K. Haghi, *Carbon Nanotube Devices for Nanoelectronics*, Synthesis Lectures on Solid State Materials and Devices,
https://doi.org/10.1007/978-3-032-02114-4

MIX
Papier aus verantwortungsvollen Quellen
Paper from responsible sources
FSC® C105338

If you have any concerns about our products,
you can contact us on
ProductSafety@springernature.com

In case Publisher is established outside the EU,
the EU authorized representative is:
Springer Nature Customer Service Center GmbH
Europaplatz 3, 69115 Heidelberg, Germany

Printed by Libri Plureos GmbH
in Hamburg, Germany